Adaptations of Desert Organisms

Edited by J. L. Cloudsley-Thompson

Volumes already published

Ecophysiology of the Camelidae and Desert Ruminants
By R.T. Wilson (1989)

Ecophysiology of Desert Arthropods and Reptiles
By J.L. Cloudsley-Thompson (1991)

Plant Nutrients in Desert Environments
By A. Day and K. Ludeke (1993)

Seed Germination in Desert Plants
By Y. Gutterman (1993)

Yitzchak Gutterman

Seed Germination
in Desert Plants

With 139 Figures

Springer-Verlag
Berlin Heidelberg New York
London Paris Tokyo
Hong Kong Barcelona
Budapest

Prof. Dr. Yitzchak Gutterman
Jacob Blaustein Institute for Desert Research
and Department of Life Sciences
Ben-Gurion University of the Negev
Sede Boker Campus 84993
Israel

Cover illustration: photograph by J.L. Cloudsley-Thompson

ISBN-13:978-3-642-75700-6   e-ISBN-13:978-3-642-75698-6
DOI: 10.1007/978-3-642-75698-6

Library of Congress Cataloging-in-Publication Data. Gutterman, Y. (Yitzchak), 1936– Seed germination in desert plants / Y. Gutterman. p. cm. – (Adaptations of desert organisms) Includes bibliographical references and index.
ISBN-13:978-3-642-75700-6
1. Desert plants – Adaptation. 2. Desert plants – Seeds. 3. Desert plants – Ecology.
4. Germination. I. Title. II. Series. QK922.G8 1993 581.5′2652–dc20 93-29023 CIP

This work is subject to copyright. All rights are reserved, whether the whole or part of the material is concerned, specifically the rights of translation, reprinting, reuse of illustrations, recitation, broadcasting, reproduction on microfilm or in any other way, and storage in data banks. Duplication of this publication or parts thereof is permitted only under the provisions of the German Copyright Law of September 9, 1965, in its current version, and permission for use must always be obtained from Springer-Verlag. Violations are liable for prosecution under the German Copyright Law.

© Springer-Verlag Berlin Heidelberg 1993
Softcover reprint of the hardcover 1st edition 1993

The use of registered names, trademarks, etc. in this publication does not imply, even in the absence of a specific statement, that such names are exempt from the relevant protective laws and regulations and therefore free for general use.

Production Editor: Herta Böning, Heidelberg
Reproduction of the figures: Gustav Dreher GmbH, Stuttgart
Typesetting: K+V Fotosatz GmbH, Beerfelden
31/3145-5 4 3 2 1 0 – Printed on acid-free paper

Preface

This book is based on the author's 30 years of research in the desert, as well as on the research of others, on the dispersal and germination of annual angiospermae, geophytes and other perennial desert plants of the Negev Desert Highlands and the other deserts of Israel and the Sinai Peninsula. The findings are compared with the results of research that has been carried out in other hot deserts of the world. The focus of the book is on the extreme and unpredictable environmental conditions which influence the germination and survival of plants in the desert. These include the wide variation in the amount and distribution of rain and the differences in the length of the growing season from one year to the next, wide fluctuations in temperature, high salinity and the pressures of seed predation. The book presents an overview of the environmental factors influencing seed germination during plant development and seed maturation, the mechanisms of seed dispersal and storage of the seed bank as well as those that affect germination during seed imbibition and seedling survival.

The survival of plants under desert conditions is connected mainly with the germination mechanisms which ensure germination and seedling development at the right time and in a suitable place. Thus, during the plant life cycle the seed has the highest resistance to extreme environmental factors, whereas the seedling has the lowest. Are there special mechanisms in some species which are able to predict the best time and place for germination? It seems that there are two main directions in which evolution has taken place:

1. Plants with survival mechanisms that will enable seeds to be dispersed and/or germinate when the chance of seedling establishment is very high and the risk relatively low. Seeds of these plants only germinate after relatively large amounts of rain and, in many cases, have relatively large seeds which are well protected against many granivores.

2. In the opposite direction of evolution are species which produce tiny seeds in very large numbers, and many of these seeds germinate after even less than 10 mm of rainfall. These are condi-

tions of high risk for seedling establishment if additional rain does not follow fairly quickly.

One of the main topics considers what is special about the dispersal and germination of the seeds of plants that are found in deserts.

The dispersal and germination strategies and mechanisms discussed focus mainly on annual plants in which germination is the most critical process before each growing season.

Limited seed separation – synaptospermy – is common among plant species inhabiting extreme deserts. Many synaptospermic species disperse their seeds by rain. At least two advantages of species with synaptospermic seed dispersal are seen particularly in extreme deserts: the retention of seeds in the favourable microhabitat where the mother plant had successfully completed its life cycle; and the protection of these seeds against seed collectors during the period from seed maturation to germination.

The patchy distribution of annuals in depressions, runnels and other more 'favourable' microhabitats, which are so pronounced in the more extreme deserts, is also discussed. This includes the succession of annuals in porcupine diggings and the appearance of different age groups in the seed bank.

The genotypic evolution of a species increases the fitness of that species for its habitat. This may increase the predictability of seeds to germinate at the proper time and place. In addition to the genotypic inheritance which causes the uniformity of seeds, phenotypic influences are also very important and result in different germinability of seeds, even those which mature on one plant, one branch or even in the same fruit. Phenotypic differences in seed germinability can be affected by environmental and maternal conditions: from the time of development of the mother plant, as a result of conditions of seed maturation; position of the seed on the mother plant, the inflorescence or fruit, as well as storage conditions and the environmental conditions during imbibition. This allows the risk to be spread as, even under optimal conditions for germination, only a portion of the seeds will germinate after one rainfall.

Acknowledgements

To my father, the late Israel David Gutterman, who encouraged me in my studies during his lifetime but did not have the opportunity to see this book. Special thanks also to my mother, Hannah Gutterman, for her help and support throughout my life; to my wife Mina and my children, as well as to my colleagues and students for their encouragement during the writing of this book.

The author is grateful for the opportunity he had to be a student and, later, a colleague of the late Prof. Michael Evenari with whom he studied in the desert for more than 20 years. It was also a great privilege to work with the late Prof. Fritz Went and to have the opportunity for long discussions on the survival mechanisms of desert plants with these two great scientists and pioneers of ecophysiological research of desert plants.

The authors wishes to thank Prof. J.L. Cloudsley-Thompson for his useful comments and careful editing of this book, and also Mrs. Frieda Gilmour.

Contents

1	The Negev Desert of Israel and Other Hot Deserts of the World: Classification According to Quantities, Distribution and Seasons of Rain. Their Influences on the Germination of Desert Plants	1
1.1	Classification of Deserts of Semi-Arid and Arid Zones and the Season of Rain	1
1.2	Abiotic Environmental Factors of the Negev Desert	2
1.2.1	Quantities and Distribution of Rain	2
1.2.2	Range of Temperatures, Relative Humidity and Dew	7
1.2.3	Species, Length of Growing Season, Environmental and Micro-Habitat Effects	10
1.2.4	Habitats and Topographic Sites	12
1.3	Some Biotic Factors Affecting the Vegetation of the Negev Desert	14
1.3.1	Overgrazing and the Activity of Seed Eaters	14
1.3.2	Porcupine Diggings as a Favourable Desert Micro-Habitat	16
1.4	Autecological Adaptations, Life Forms and the Annual Cycle of Desert Angiospermae	17
1.4.1	Arido-Passive Perennials and Annuals	18
1.4.2	Arido-Active Plants	19
1.4.3	Species, Ecogenotypes and Ecotypes	20
1.4.4	Genotypic and Phenotypic Influences on Seed Germination	23
2	Phenotypic Effects on Seeds During Development	25
2.1	Introduction	25
2.2	Position Effects	26
2.2.1	Position Effects in Plants with Aerial and Subterranian Fruits (Amphicarpy)	26
2.2.2	Position of the Capsules on the Plant Canopy Affects Germination	28

2.2.3	Position Effects of Seeds in the Inflorescence When It Is a Dispersal Unit	29
2.2.4	Position of Achenes in the Capitulum Whorls and Their Germination	36
2.2.5	Position of Female and Hermaphroditic Flowers and Their Seed Germinability	36
2.2.6	Position of Flowers and Dimorphism	37
2.2.7	Position, Heteromorphism and Germinability	39
2.2.8	Position Effect of Seeds in the Fruit	39
2.2.9	Position and Annual Rhythm in Seed Germinability	40
2.3	Age Effects	41
2.3.1	Senescence of the Mother Plant Affecting Seed Germination	41
2.3.2	Age of the Mother Plants Affecting Seed Germination Was also Found in Non-Desert Plants	41
2.4	Environmental Effects	41
2.4.1	Photothermal and Position Effects	41
2.4.2	Day Length Affecting Seed Germination of Plant Species with Dry Fruit	42
2.4.3	Day Length Affecting Seed Germination of Species with Soft Fruit	53
2.4.4	The Critical Time for the Day Length Effects	56
2.4.5	Natural Conditions of Maturation During Summer and Autumn Affecting Seed Germination	57
2.4.6	Natural Winter, Spring or Summer Conditions of Maturation Affecting Seed Germinability	60
2.4.7	Altitudinal Effects	61
2.4.8	Influences of Light Quality During Maturation on Seed Germination	62
2.4.9	Temperatures During Maturation Affecting Seed Germination	70
2.4.10	Conditions of Maturation and Different Levels of Germination Under Different Temperatures During Imbibition	71
2.4.11	Water Stress During Maturation Affecting Seed Germination	71
2.4.12	Dimorphism and Achene Germination Dependent on Plant Size	73
2.4.13	Heterocarpy and Germinability	73
2.5	Polymorphic Seeds and Germination	74
2.6	Conclusions	74

Contents XI

3	Seed Dispersal and Seed Predation of Plants Species in the Negev Desert	79
3.1	The Annual Cycle of Seed Maturation and Dispersal of Some of the Common Species in the Negev Desert Highlands	79
3.2	Seed Maturation and Dispersal in Summer	81
3.2.1	Seed Dispersal by Wind	82
3.2.2	Porcupine Diggings and Other Depressions as Wind-Traps for Seeds, and the 'Treasure Effect' .	93
3.2.3	Seed Predation and Seed Dispersal by Ants, Birds and Mammals	95
3.2.4	Changes in the 'Seed Bank' – During Summer .	100
3.2.5	Seed Dynamics in the Soil of Species Dispersed by Wind	102
3.2.6	Changes in the 'Seed Bank' of Plants That Disperse Their Seeds by Rain	103
3.3	Seed Maturation in Summer and Dispersal by Rain in Winter	105
3.3.1	Plants Whose Seeds Are Dispersed by Rain (Ombrohydrochory) and/or Runoff Water	105
3.3.2	Mechanisms of Seed Release by Rain	105
3.3.3	Differences in Dispersal Mechanisms of Some Groups of Species	112
3.4	Seed Maturation in Summer and Germination in Situ (Atelechory) in Winter	138
3.5	Seeds That Mature and Are Dispersed in Winter	138
3.6	Heterocarpy and Species Survival	139
3.7	Conclusions	140
4	Storage Conditions Affecting the Germinability of Seeds in the Seed Bank	145
4.1	Introduction	145
4.1.1	Seed Bank Location of Some Desert Plant Species	145
4.1.2	Soil and Seed Turnover	147
4.1.3	Environmental Factors and the Changes at Different Soil Depths	149
4.2	Seed Internal and Environmental Factors During Storage Affecting Germination	151
4.2.1	Seed Coat Effects	151
4.2.2	Short, Wet Storage at High Temperatures	158
4.2.3	The Short-Term Changes of Seeds Influenced by Temperatures During Dry Storage	158
4.2.4	Long-Term Changes During Dry Storage and Viability	161
4.3	Conclusions	165

4.3.1	Short-Storage Seeds	165
4.3.2	Long-Storage Seeds and Regulation of Germination	166
4.3.3	Seeds with Strategies of Dispersal and Germination over Many Years	166
5	Environmental Factors During Seed Imbibition Affecting Germination	169
5.1	Introduction	169
5.1.1	Germination at the Right Time	169
5.1.2	The Genotypic and Phenotypic Regulation for Seed Germination	170
5.1.3	High- or Low-Risk Strategies for Seed Dispersal and Germination	171
5.2	Water	175
5.2.1	The Minimum Amount of Rain That Engenders Germination in Desert Plants?	175
5.2.2	Germination Inhibitors as 'Rain Gauges'	180
5.3	The Period of Wetting for Germination	186
5.3.1	Fast-Germinating Seeds	188
5.3.2	Slow-Germinating Seeds	188
5.3.3	Air, Mucilaginous Seeds (Myxospermy) and Seed Germination	189
5.4	Temperature	189
5.4.1	Temperature and Germination of Winter or Summer Annuals	190
5.4.2	Temperatures, Day Length and Seasonal Genoecotypes	191
5.4.3	Thermodormancy and Winter-Germinating Species	191
5.4.4	Summer-Germinating Species	192
5.4.5	Different Strategies of Two Negev Plants	193
5.4.6	Species Habitat, Speed and Range of Temperatures for Germination	194
5.4.7	Thermoperiodism Affecting Seed Germination	195
5.5	Light and Germination	198
5.5.1	Maturation in Natural Conditions Affecting Seed Light Sensitivity	199
5.5.2	Maturation under Artificial Light Affecting Seed Light Sensitivity	199
5.5.3	Storage Conditions and Seed Light Sensitivity	200
5.5.4	Types of Response of Seeds to Light	200
5.6	Annual Rhythm Regulating Seed Germination	204
5.7	Mass Germination and Seedling Emergence from Below the Soil Crust	205
5.8	Conclusion	206

Contents XIII

6	Germination, the Survival of Seedlings and Competition	207
6.1	Introduction	207
6.1.1	Selective Process of Seed Germination and Seedling Survival	207
6.1.2	Rain Amount, Distribution and Survival	207
6.1.3	Micro-Habitat and Seedling Survival	210
6.1.4	Day Length and Water Stress Affecting Life Span of Annuals	210
6.2	'Opportunistic' or 'Cautious' Strategy. Low or High Chance of Seedling Survival	215
6.2.1	Seedling Congestion and Survival	215
6.3	Depressions and Porcupine Diggings as Favourable Micro-Habitats in the Desert	218
6.4	Drought Tolerance and the Survival of Seedlings	220
6.4.1	Seedling 'Point of No Return'	220
6.4.2	Plant Drought Tolerance	221
6.5	Mass Germination and Age Groups Replace Dead Plants	221
6.6	Conclusion	222
7	Conclusion	225
References		231
Latin Name Index		247
Subject Index		251

1 The Negev Desert of Israel
and Other Hot Deserts of the World:
Classification According to Quantities, Distribution
and Seasons of Rain.
Their Influences on the Germination of Desert Plants

1.1 Classification of Deserts of Semi-Arid and Arid Zones and the Season of Rain

According to the average amount of annual rainfall Zohary (1962), followed by Blair (1942), classified Israel into three vegetative zones: (1) 1000–400 mm – sub-humid zone; (2) 400–200 mm – semi-arid zone; (3) 200–25 mm – arid zone. These zones coincide more or less with the plant geographical divisions. The Negev is in the third category of the arid zone of Israel, with Beer Sheva in the north (31°14'N, 34°48'E) receiving 200 mm of rain, Sede Boker (30°51'N, 34°46'E) 100 mm and, in the south, Eilat (29°33'N, 34°57'E) receiving 25 mm.

The Negev is important both from the point of view of plant distribution and of illustrating the survival mechanisms of plants under unpredictable and extreme desert conditions. First, the Negev is part of the belt of hot deserts of North Africa and the Middle East. In addition, the Negev desert is situated at the meeting point of four very large geographical areas of plant distribution: the Saharo-Arabian, Irano-Turanian, Mediterranean and the Sudanian areas. The Negev and the Jordan Valley are the southern point of continuous distribution for the Irano-Turanian and Mediterranean plants and the northern point of distribution for the Sudanian and the Saharo-Arabian plants (Zohary 1962, 1966, 1972; Feinbrun-Dothan 1978, 1986).

Not only do the annual amount and distribution of rain affect the germination and development of plants, but not less significant is whether the rain appears in the hot or in the cold season of the year or in both seasons. When soils taken from a desert that receives both winter and summer rain were wetted (Went 1948, 1949, 1957), seedlings of winter annuals appeared under low temperatures, while seedlings of summer annuals appeared at high temperatures.

The northern part of the Sahara and the Saudi Arabian desert receives winter rain (Kassas 1966; Batanouny 1981), the southern part summer rain, whereas the central area receives very little infrequent rain. Similar phenomena are also characteristic of the Karoo desert where the western part receives rain during the cool season and the eastern part during the summer (Werger and Coetzee 1978). Of the deserts of North America, the Mojave receives 50–100% of the 30–300 mm rain in winter. The Chihuahua only receives up to 40% in winter of the 180–450 mm annual precipitation. The Sonoran desert receives in winter 35–65% of the annual 50–300 mm of rain (Mac-Mahon and Wagner 1985). The question arises, what are the mechanisms of

2   The Negev Desert of Israel and Other Hot Deserts of the World

seed germination that ensure the germination of each plant species in the right place and at the right time of the year?

1.2 Abiotic Environmental Factors of the Negev Desert

1.2.1 Quantities and Distribution of Rain

A summary of 15 years of rainfall measurements at the Desert Research Institute at Sede Boker (Zangvil and Druian 1983), plus knowledge obtained from field observations in the Avdat region of the Negev highlands (Evenari and Gutterman 1976) indicate unpredictable and very large fluctuations in the quantity of rain and in its distribution from one year to another.

1.2.1.1 The Length of the Rainy Season, Amounts and Distribution of Precipitation

1. The onset of the seasonal rains at Sede Boker (0.1 – 16.3 mm, average 2.6; Table 1), or in Avdat (1.5 – 9.4 mm, average 5.3 mm; Table 2), following the

Table 1. First and last rains of each season from 1976/77 to 1990/91, length of season between the first and last rain, total number of days with rain and total amounts of rain (mm) per season. (After observations of rainfall at the Unit of Meteorology, Jacob Blaustein Institute for Desert Research, Ben-Gurion University of the Negev, Sede Boker)

Year	First rain		Last rain		No. of days between first and last rain per season	No. of days with rain	Total mm rainfall per season
	Date	Amount	Date	Amount			
1976/77	13. X. 76	0.1	13. V. 77	4.2	213	26	86.2
1977/78	17. X. 77	0.6	4. VI. 78	0.3	230	23	53.0
1978/79	15. X. 78	0.9	4. V. 79	0.6	209	25	74.9
1979/80	20. X. 79	0.1	12. V. 80	0.9	205	43	159.1
1980/81	10. XII. 80	16.3	14. IV. 81	1.6	126	18	130.9
1981/82	5. XI. 81	0.6	12. V. 82	0.6	189	33	78.6
1982/83	28. IX. 82	1.5	18. IV. 83	0.6	203	42	140.9
1983/84	7. XII. 83	1.4	26. III. 84	0.2	109	22	56.4
1984/85	18. X. 84	5.0	11. V. 85	0.4	216	28	73.8
1985/86	30. X. 85	0.5	3. V. 86	4.0	186	26	146.7
1986/87	2. XI. 86	3.3	24. III. 87	4.6	142	24	75.4
1987/88	21. X. 87	2.2	22. IV. 88	1.8	182	34	106.95
1988/89	13. XII. 88	2.0	27. III. 89	drops	104	20	102.15
1989/90	6. XI. 89	0.9	2. IV. 90	2.7	147	24	93.8
1990/91	20. X. 90	3.8	23. III. 91	16.2	155	19	141.1
Minimum	–	0.1	–	0.2	104	18	53.0
Maximum	–	16.3	–	16.2	230	43	159.1
Average	–	2.6 ± 1[a]	–	2.6 ± 1.1	174.4 ± 10.5	27.1 ± 2	101.3 ± 8.9

[a] ± SE

Abiotic Environmental Factors of the Negev Desert

Table 2. First and last rains of each season from 1960/61 to 1964/65, length of season between the first and last rain, total number of days with rain and total amounts of rain (mm) per season. (After Evenari and Gutterman 1976)

Year	First rain		Last rain		No. of days between first and last rain per season	No. of days with rain	Total mm rainfall per season
	Date	Amount	Date	Amount			
1960/61	2. XI. 60	8.4	7. IV. 61	0.5	157	18	70.4
1961/62	27. X. 61	9.4	24. IV. 62	5.8	179	20	64.7
1962/63	17. I. 63	1.8	6. V. 63	1.8	109	10	29.5
1963/64	21. X. 63	1.5	21. IV. 64	3.0	182	25	165.0
1964/65	17. XI. 64	5.6	18. IV. 65	3.3	153	28	159.8
Minimum	–	1.5	–	0.5	109	10	29.5
Maximum	–	9.4	–	5.8	182	28	165.0
Average	–	5.3	–	2.9	156	20.2	97.9

hot and dry summer, may be from October in one year to December in another, at Sede Boker (Table 1), or even January at Avdat (Table 2). This means that there can be a difference of 3 to 4 months between the first rainfall of one year and that of another.

2. The last rain in the season at Sede Boker (0.2–16.2 mm, average 2.6 mm – Table 1), or at Avdat (0.5–5.8 mm, average 2.9 mm – Table 2), may occur in March or April of one year or in May of another. This is a difference of 2 to 3 months.
3. From the first rain of one year to the last in another, there can be a range of 8 months between October and May (Tables 1, 2; Figs. 1, 2, 3).
4. The number of days with rain in any one season may be from 10 in one year to 43 in another (Tables 1, 2).
5. The number of days between the first day of rain to the last in any one season can be from 104 to 230 days (Tables 1, 2).
6. The amount of rain per season ranges from about 29.5 mm to 165 mm (Table 2; Fig. 4).

1.2.1.2 Rain and Germination

1. When the proper range of temperature and relative humidity exists, the beginning of germination, after the first rain of more than 12 mm, can be from November in one year to April in another. This means that there can be a difference of 4.5 months between the first germination in one year and that of another (Tables 3, 4).
2. The last rain of more than 12 mm in the season, which might engender germination if suitable temperatures and relative humidity persist, could be in December (25th) in one year and in April (8th) in another. This is a difference of 3.5 months from one year to another.

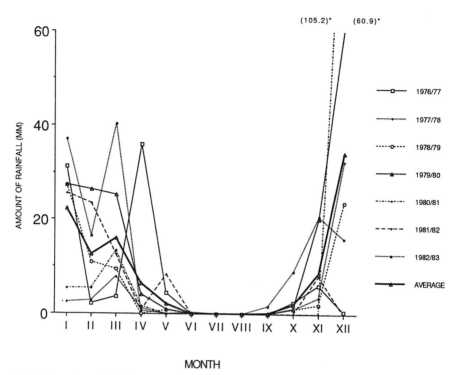

Fig. 1. Summary of rainfall (mm) per month for the years 1977–1982 at Sede Boker. (After Zangvil and Druian 1983)

3. From the first germination in one year to the last of another, there can be a range of 6 months (from October to April) (Tables 3, 4, 5).
4. The number of precipitations of at least 12 mm rainfall that could cause germination in one season in one year may range between one and five (Tables 3, 4, 5).
5. The longest consecutive period in which more than 12 mm of rain accumulates may be from 1 to 7 days (Tables 3, 4). There can also be two or more falls of rain with one or more dry days in between, which may also affect the germination of the seeds of species, such as *Artemisia sieberi*, that need long periods of imbibition before germination (Sects. 3.3, 5.3).
6. The loess soils are covered with a crust, and 70% of rainfall may be lost through runoff and evaporation. Chalk, marl and clay, especially, are covered with a biological crust. Only low amounts (30–50%) of rain penetrate the soil, and are soon lost by evaporation, which increases salinity. The runoff water accumulates in depressions (Danin 1983). Rain will, therefore, induce germination mainly in depressions, runnels, cracks in the soil crust, as well as in wadis. Runoff water accumulated in depressions, such as porcupine diggings, even after a rainfall of as little as 7.1 mm on 12–13 November 1981 and was sufficient to engender germination (Gutter-

Abiotic Environmental Factors of the Negev Desert

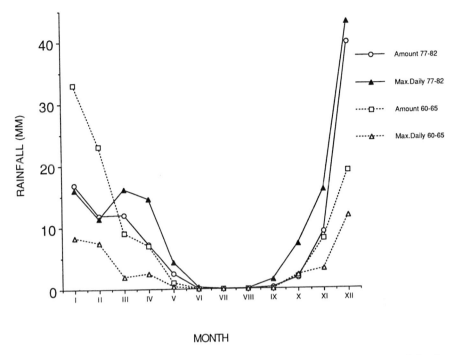

Fig. 2. Summary of yearly rainfall (mm) per month or maximum daily (mm) per month for the years 1960–1965 at Avdat (After Evenari and Gutterman 1976) and 1977–1982 at Sede Boker. (After Zangvil and Druian 1983)

man 1982a). Nearly all the water penetrates into sand. Stable-sand habitats in deserts, therefore, have a higher carrying capacity than other soils.

7. When rainfall intensity is lower, less runoff develops and more water penetrates loess soil. This may induce germination along cracks and on the flat soil surface, not only in depressions. But in depressions, such as porcupine diggings, annuals still germinate, and more seedlings develop than in surrounding slopes (Fig. 137) (Gutterman 1989a; Gutterman et al. 1990). Wadis with loess soil form the principal habitat of the Negev highlands in which annual plants germinate and survive.

1.2.1.3 The Year-to-Year Fluctuations of Amounts of Rain

Unpredictable and large fluctuations in the amounts of rain from one year to another are typical of the Negev highlands. In 1977/78, 53 mm of rain fell; in 1979/80, 159.1 mm; in 1983/84, 56.4 mm; in 1985/86, 146.7 mm (Table 1, Fig. 4). The most dramatic precipitations during the last 19 years were the 29.5 mm in Avdat in 1962/63 and 1963/64, with 165 mm a year later (Table 2) (Evenari and Gutterman 1976).

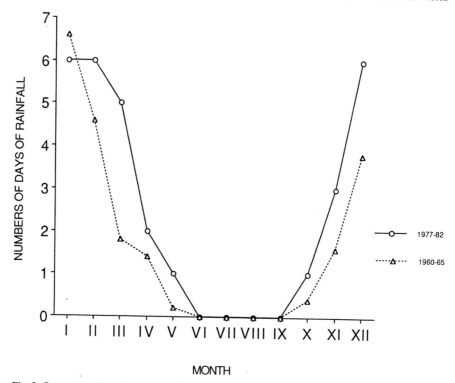

Fig. 3. Summary of number of days of rainfall per year for the years 1960–1965 at Avdat and 1977–1982 at Sede Boker. (After Evenari and Gutterman 1976; Zangvil and Druian 1983)

Such large fluctuations, which are typical of this region, together with the unpredictability of the appearance of the first and last rain of the season, make the measurement of a yearly average valueless. Annual plants, with their survival mechanisms, appear in very large numbers in years of rainfall above the average. They appear in very low numbers when the amount of rain is far below the yearly average. Perennial plants also exhibit very high flexibility. They can reduce the leaf area or area of transpiration depending on environmental factors (Evenari et al. 1982). Shrubs that survive for many decades, or even for hundreds of years, serve as meteorological stations indicating the minimum conditions for survival of this particular species.

In the east-southern part of the Negev, which usually receives below 70 mm, the average precipitation per year, nearly all the shrubs and trees appear in areas receiving additional amounts of water. These include wadis, aquifers and places with a water-table near to the soil surface. This phenomenon is also dependent on the type of soil, not only on the average rainfall (Danin et al. 1975). Only annuals appear on flat areas and these depend on seasonal amounts of rain. Such a region is regarded as extreme desert according to the classification of aridity. This takes into consideration both the average amount

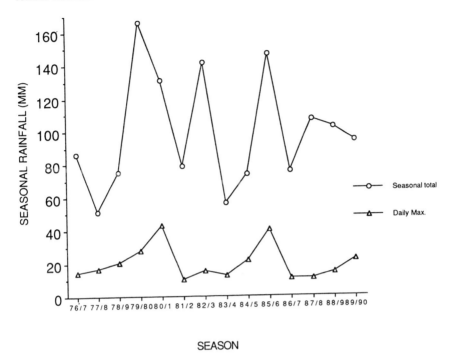

Fig. 4. Seasonal total as well as maximum daily rainfall for the years 1976–1990 at Sede Boker. (After Zangvil and Druian 1983 and the Unit of Meteorology)

of precipitation and the mean annual temperatures (Köppen 1954; Rudloff 1981; Griffiths and Driscoll 1982). Aridity is also dependent on the type, topography, elevation and salinity of the soil. Among annual plants which display opportunistic behaviour, the amounts of water that penetrate the soil and the distribution of the seasonal rain are together major factors for germination and the establishment of the seedlings.

1.2.2 Range of Temperatures, Relative Humidity and Dew

As can be seen from Fig. 5, the hottest months of the year are July and August. The lowest temperatures of the year occur in December and January with a daily minimum of 4 °C, an average of 10 °C and a daily maximum of 15 °C (Fig. 5). These are suitable temperatures for germination of the plant species that have been studied in this area. December and January are also the months with the minimum evaporation rates (60 mm per month). The maximum is in June and July (300 mm per month) (Figs. 6, 7). The relative humidity is at its lowest level (about 27.5%) in June and highest in December (about 55%) on a daily average (Fig. 8).

Table 3. Summary of rainfall events at Sede Boker in the Negev Desert highlands, from 1976–1991. Dates and amounts of rain of over 12 mm per one or two consecutive dates and longest periods of more than 12 mm of rain in one rain event per season, and the total rainfall per season. (After observations on the rainfall at the Unit of Meteorology, Jacob Blaustein Institute for Desert Research, Ben-Gurion University of the Negev, Sede Boker)

Year	Rain of more than 12 mm per 1 or 2 consecutive days					Longest period of continuous rain of more than 12 mm			Total mm rainfall per season
	First rain		Last rain		Number of rain events	No. of days	first day	total mm	
	Date	Amount	Date	Amount					
1976/77	12. IV. 77	15.1[a]	–	0	1	7 days	2. I. 77	23.5	86.2
1977/78	22. XII. 77	23.3[a]	–	0	1	4 days	21. XII. 77	23.6	53.0
1978/79	11. XII. 78	21.7[a]	9. I. 79	19.7	2	5 days	8. XII. 78	23.2	74.9
1979/80	29. XI. 79	19.9[c]	3. III. 80	17.6	5	4 days	28. XI. 79	20.4	159.1
1980/81	10. XII. 80	47.0	25. XII. 80	53.7	2	3 days	25. III 81	13.3	130.9
1981/82	26. I. 82	(10.2)[b]	–	0	0	4 days	25. I. 82	14.8	78.6
1982/83	23. I. 83	13.3	4. III. 83	25.7	2	5 days	7. XI. 82	17.6	140.9
1983/84	15. I. 84	13.5[a]	–	0	1	4 days	13. III. 84	13.2	56.4
1984/85	14. II. 85	12.2	22. III. 85	21.8[a]	2	2 days	14. II. 85	12.2	73.8
1985/86	17. XII. 85	19.2	8. IV. 86	51.1	2	3 days	17. XII. 85	41.9	146.7
						3 days	1. IV. 86	15.7	
1986/87	28. XI. 87	15.7	–	0	1	6 days	27. XI. 86	19.2	75.4
1987/88	15. I. 88	14.0	3. III. 88	17.3	3	5 days	3. I. 88	15.5	106.95
						4 days	15. I. 88	17.8	
1988/89	23. XII. 88	21.5[a]	11. II. 89	14.0	3	6 days	23. XII. 88	38.3	102.15
1989/90	26. I. 90	22.3[a]	1. IV. 90	21.6	3	2 days	1. III. 90	20.2	93.8
1990/91	25. I. 91	53.3[a]	23. III. 91	16.2	4	3 days	24. I. 91	54.1	141.1
						3 days	21. III. 91	47.0	

[a] Heaviest rain per season.
[b] In ( ) heaviest rain per season less than 12 mm.
[c] The heaviest rain this season was 27.6 mm on 14. XII. 79.

Table 4. Summary of rainfall events at Avdat in the Negev Desert highlands, from 1960/61 – 1964/65; dates and amounts of rain of over 12 mm per one or two consecutive dates and longest periods of more than 12 mm of rain in one rain event per season, and the total rainfall per season. (After Evenari and Gutterman 1976)

Year	Rain of more than 12 mm per 1 or 2 consecutive days				Number of rain events	Longest period of continuous rain of more than 12 mm			Total mm rainfall per season
	First rain		Last rain			No. of days	first day	total mm	
	Date	Amount	Date	Amount					
1960/61	1. II. 61	13.0	16. II. 61	13.2	2	1	1. II. 61	13.0	
						1	16. II. 61	13.2	70.4
1961/62	6. XII. 61	14.7	–	0	1	1	6. XII. 61	14.7	64.7
1962/63	10. II. 63	13.5	–	0	1	2	10. II. 63	13.5	29.5
1963/64	1. XII. 63	41.6	18. I. 64	13.7	5	4	1. XII. 63	64.1	
						4	30. XII. 63	31.7	165.0
1964/65	12. XII. 64	15.5	18. I. 65	12.0	4	6	8. I. 65	59.6	159.8

Table 5. Interval between the first and last rain of more than 12 mm and the number of such events, per season. (After observations of the rainfall at the Unit of Meteorology, Jacob Blaustein Institute for Desert Research, Ben-Gurion University of the Negev, Sede Boker)

Year	The first month to the last month with rain events of more than 12 mm in one season[a]	Number of rain events of more than 12 mm per season
1976/77	January – April	2
1977/78	October – December	1
1978/79	December – January	2
1979/80	November – March	5
1980/81	December – March	3
1981/82	January	1
1982/83	November – March	3
1983/84	January – March	2
1984/85	February – March	2
1985/86	December – April	3
1986/87	November	1
1987/88	January – March	4
1988/89	December – February	3
1989/90	January – April	3
1990/91	January – March	3

[a] In the deserts of Israel there are mainly two seasons: (1) the season with rain and low temperatures, and (2) the dry and hot season.

The long-term annual total amount of dew is 17 mm in 192 nights during 1377 h. Maximum levels of dew were measured in December and January with a lower peak in April (Fig. 9). Almost the same pattern was observed in the number of hours of dew per month as well as the number of nights with dew per month. Only 0.6 mm in 53 h of dew were measured during 8 nights in April; in January, there were 2.1 mm of dew in 171 h during 18 nights and in September 1.9 mm in 157 h during 25 nights (Zangvil and Druian 1983) (Fig. 10a, b).

1.2.3 Species, Length of Growing Season, Environmental and Micro-Habitat Effects

The environmental factors already mentioned, together with others such as the amounts of radiation, day length and water-stress (Gutterman 1982b), are all involved in the length of the growing season of the plants of this area. There are large differences between one species and another, between one type of adaptation and another, as well as differences according to the soil type, micro-topography, slope direction and other conditions.

The critical time for starting the growing season of annuals is the first rain that causes germination. Temperature, water-stress and day length are the main factors that induce the seeds of some winter annual species to mature and the

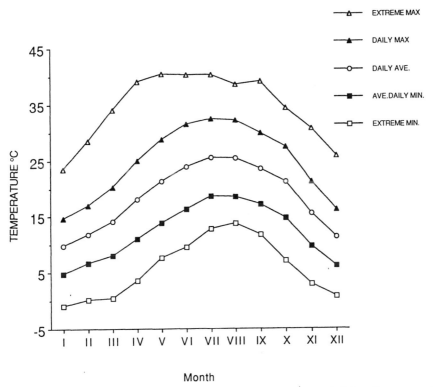

Fig. 5. Average monthly temperature data for the years 1977–1982 at Sede Boker: Extreme maximum, daily maximum, daily average, average daily minimum, extreme minimum. (After Zangvil and Druian 1983)

plants to dry up at the beginning of the summer (Evenari and Gutterman 1966; Gutterman 1982a, b, 1989c, d, e, f). Species of perennials that have root systems near the soil surface, such as *Carex pachystylis*, are the first to begin growing after relatively small amounts of rain (Evenari et al. 1982). For geophytes, there are various species with differing strategies reflected in the depth of the bulbs, corms and rhizomes, etc. According to the depth of their storage organs, different species begin to develop roots when soil moisture reaches the area of root initiation. *Bellevalia desertorum*, of which the bulb is situated only about 5 cm below the soil surface, begins its growing season much earlier than its relative, *B. eigii*, the bulb of which is situated about 25 cm below the soil surface. *B. desertorum* flowers, produces seeds and its leaves dry out much earlier than *B. eigii*, even when both species inhabit the same wadi bed (Boeken and Gutterman 1991). The seeds of *B. eigii* germinate near the soil surface and the developing bulbs change their strategies according to the increase in the depth of the bulb over the years.

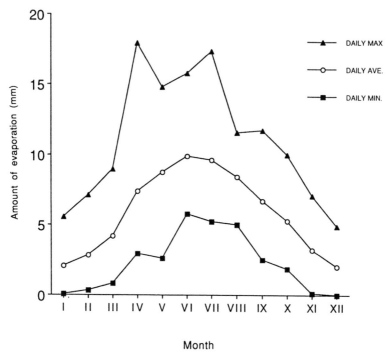

Fig. 6. Summary of daily evaporation (mm), daily average and daily minimum, per month from water-free space type "A" evaporation pan for years 1977–1982 at Sede Boker. (After Zangvil and Druian 1983)

1.2.4 Habitats and Topographic Sites

In extreme hot desert habitats, the season suitable for germination and growth is short and unpredictable (Tables 1–5; Figs. 1, 4). In the Negev the loess soil allows water penetration of 30–50%. Under such conditions, run-off water develops even after a small amount of rain (3–4 mm/h). In this type of soil, run-off water accumulates at sporadic intervals in depressions, such as porcupine diggings, during winter. These micro-sites are favourable for seed germination and seedling development (Fig. 71) (Gutterman and Herr 1981; Gutterman 1989a; Gutterman et al. 1990) (Sects. 3.2.2, 4.1.2, 5.2.1, 6.3).

Only about 30% of the precipitation that falls in a given slope area enters the soil surface between furrows. In contrast, in diggings and depressions, run-off water accumulates, and the amount of water there can be several times more than the average amount of precipitation in the area. Plants that finish their life cycle in such furrows or diggings benefit greatly. The same microhabitat will be inhabited by the same species in subsequent winters when they have an atelechoric mechanism for seed dispersal (Zohary 1937; Gutterman 1980/81a, 1990b; Gutterman et al. 1990) (Sect. 3.2.2).

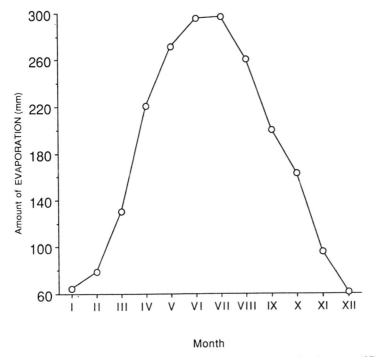

Fig. 7. Summary of amount of evaporation (mm) per month for the years 1977–1982 at Sede Boker. (After Zangvil and Druian 1983)

The water input of loess soil also depends upon macro-scale topography of the surface of the site since this influences the run-off/run-on gradient (Beatley 1974). According to Evenari et al. (1971, 1982), there are nine main habitats in the Negev desert. In a year with about 100–130 mm rain, these habitats will receive the following amounts of rain:

1. The plateaux on the hills will absorb 10–20 mm to a depth of about 20 cm.
2. The hill slopes will absorb 20–50 mm to a depth of about 50 cm.
3. Wadi runnels on a loess plain will absorb 200 mm to a depth of about 200 cm.
4. A loessial plain will receive 30 mm to a depth of about 30 cm.
5. A large wadi with a wadi bed of deep loess soil will receive 500–600 mm to a depth of about 500 cm.
6. A large wadi with a wadi bed of gravel and stones will receive 80–100 mm to a depth of about 700 cm.

These six habitats (Fig. 11) (Evenari et al. 1971) cover about 90% of the Negev, and there are three more types of habitat:

7. Sand fields with no or a very poor run-off which will receive 100–130 mm.

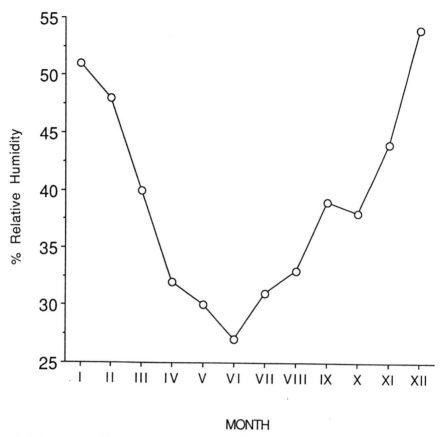

Fig. 8. Summary of daily average of relative humidity % for each month for the years 1977–1982 at Sede Boker. (After Zangvil and Druian 1983)

8. Saline areas in which water accumulates above the soil surface and evaporates.
9. Oases which receive their water from springs or floods.

There are still more habitats, such as soil belts and soil pockets on rocky slopes, which are inhabited by species or ecotypes of the Mediterranean zone, as well as by species that inhabit wadis in this area (Danin 1983).

1.3 Some Biotic Factors Affecting the Vegetation of the Negev Desert

1.3.1 Overgrazing and the Activity of Seed Eaters

Continuous overgrazing, over thousands of years, has had a very large influence on the vegetation of the Negev highlands. Many species of grasses have

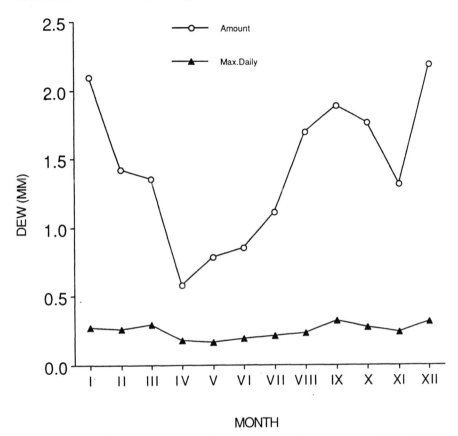

Fig. 9. Summary of dew (mm) per month, or maximum daily dew per month for the years 1977–1982 at Sede Boker. (After Zangvil and Druian 1983)

disappeared over large areas and can only be found in limited and hidden places. Other influences, especially on the changes of the 'seed banks', are the seed consumers — mainly ants (Evenari 1981) but also birds and rodents (Bar et al. 1984). It has been found that, where more seeds accumulate in a depression or digging, this is more attractive for seed consumers, and fewer seeds remain until the following rainy season — the 'treasure effect' (Sect. 3.2.2) (Gutterman 1988a).

Many plant species protect their seeds and disperse them during the following rainy season (Gutterman 1990a). Some protect their seeds in other ways, such as by subterranean achenes or propagules which are protected by the dead mother plants — the dead serving the living. These atelechoric 'non-dispersal' fruits are protected from seed eaters in a micro-habitat where the probability for the seedling to survive is much higher than for the seeds of telechoric aerial dispersal units of the same plant species (Koller and Roth 1964; Evenari et al. 1977; Evenari 1981). The seeds of other plants, such as *Artemisia sieberi*,

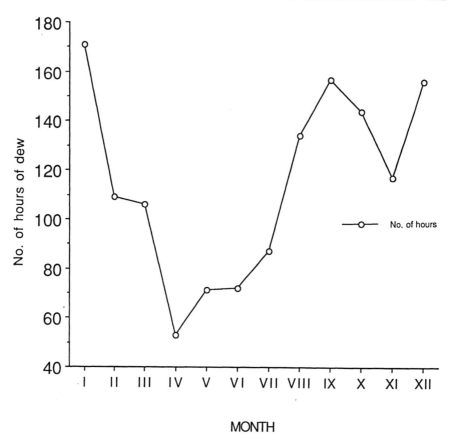

Fig. 10. a Summary of number of hours of dew per month for the years 1977–1982 at Sede Boker. (After Zangvil and Druian 1983)

mature in the rainy season, when the time of seed maturation is close to the time of seed germination (Sects. 2.2, 3.5).

1.3.2 Porcupine Diggings as a Favourable Desert Micro-Habitat

Porcupine (*Hystrix indica*) diggings, left behind after consumption of the storage organs of geophytes and hemicryptophytes, have a large influence on the vegetation of the Negev highlands. These diggings are a very favourable habitat for the germination of annuals and the renewal of geophytes and hemicryptophytes. In certain areas there may be as many as two diggings per m^2. The system can be seen as one which opens closed spaces of disturbance and recovery, in which even a succession of annuals has been observed. In some areas, almost all the soil surface is covered with porcupine diggings in various stages of recovery (Gutterman 1982c, 1987, 1988a, 1989a; Gutterman

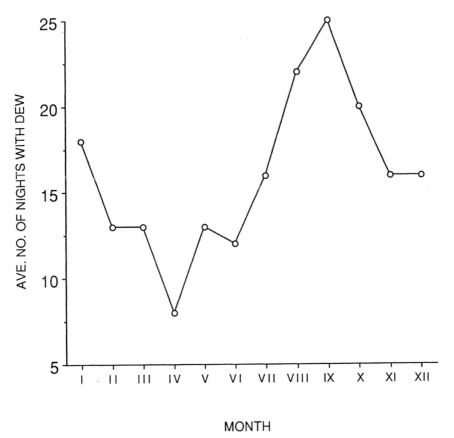

Fig. 10b. Summary of monthly average of nights with dew for the years 1977–1982 at Sede Boker. (After Zangvil and Druian 1983)

et al. 1990). (For more details see Sect. 3.2.2, Fig. 71; Sect. 4.1.2, Fig. 113a, b; Sect. 6.3, Figs. 137, 138).

1.4 Autecological adaptations, life forms and the annual cycle of desert angiospermae

According to their autecological adaptations and life forms, the higher plants of the Negev can be divided into two main groups: the arido-active plants which remain active throughout the year, and the arido-passive plants which are not, or are much less, active during the hot and dry season (Evenari et al. 1971, 1982; Evenari and Gutterman 1976, 1985; Gutterman 1981).

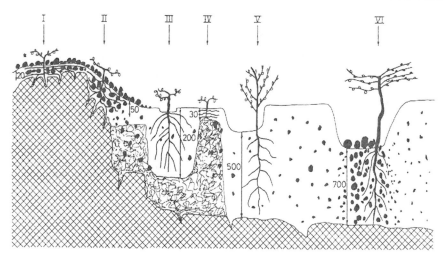

Fig. 11. Schematic diagram (not to scale) of six main habitats of the Negev highlands: *I* hilltop; *II* slope; *III* depression or runnel in loess plain; *IV* loess plain; *V* loess wadi; *VI* gravel wadi. *Arabic figures* indicate the depth (cm) of the wetted soil profile; *cross hatching*, parent rock; *vermiculation*, permadry soil; *black spots*, stones. (After Evenari et al. 1971)

1.4.1 Arido-Passive Perennials and Annuals

Winter annuals appear during the short period when there are relatively suitable environmental conditions for seed germination and development. Plants such as *Schismus arabicus* (Fig. 54) and *Stipa capensis* complete their life cycle by producing and dispersing mature seeds by the beginning of the hot and dry season. The seeds remain as a 'seed bank' in the soil or amidst the dry remains of dead mother plants. In other species, seeds remain as a 'seed bank' protected by the dead mother plant and are dispersed over many years. Examples of the latter include *Blepharis* spp., *Asteriscus pygmaeus* and *Trigonella stellata* (Gutterman et al. 1967; Evenari and Gutterman 1976; Gutterman 1982a, 1990b) (see Chaps. 2, 3, 4, 5).

Synanthus geophytes are plants in which the roots develop and the leaf canopy begins to appear above the surface at the beginning of the rainy season, only after the soil has become wet in the area of the root initiations on the bulbs, corms or rhizomes (Gutterman 1981). After flowering in the early spring, and seed maturation, the root system, leaves and fruits dry out at the beginning of summer. The subterranean storage organ remains until the following rainy season (Danin 1983; Evenari and Gutterman 1985). Examples are *Bellevalia desertorum*, *B. eigii* (Boeken and Gutterman 1989a, 1991), *Tulipa systola* (Gutterman 1981; Boeken and Gutterman 1989b), or *Ornithogalum trichophyllum* (Fig. 12d) and *Carex pachystylis* (Fig. 12f) (Evenari et al. 1971, 1982).

It is interesting to compare the life cycle of the Irano-Turanian synanthus geophytes of the Negev Desert highlands, which is the southernmost edge of distribution of these species, with that of similar species inhabiting the cool Irano-Turanian deserts of central Asia. The appearance and disappearance of the root system of bulb species of *Allium* are dependent, in both areas, upon the wetting of the soil at the beginning of the autumn rainy season to the depth of root initiation, and drying in summer. The monocarpic shoots begin to grow shortly before or at the beginning of the season, as in the Negev; however in the cool deserts they stop growing when still below the surface until the snow covering the surface has melted in the spring. Once the snow has melted, and the soil temperature has risen, the monocarpic shoots develop quickly and the leaf canopy appears above ground. They flower in May-June (Baitulin et al. 1986; Kamenetsky 1987, 1988).

Hemicryptophytes are plants with a very similar life cycle to that of geophytes, but the renewal buds are situated close to the soil surface and they have one or more main storage roots, or cormlets. They are active during the rainy season and are without leaves during the hot, dry season. Examples include *Scorzonera papposa* (Fig. 59), *Scorzonera judaica* and *Erodium crassifolium* (Fig. 12b) (Evenari et al. 1971; Gutterman 1981; Danin 1983).

1.4.2 Arido-Active Plants

Bi-Seasonal Annuals. There are some species, such as *Salsola inermis* (Fig. 12a) (Evenari et al. 1971) which germinate in winter and develop into a rosette-like stage in which they remain until the beginning of summer. When the seeds of winter annuals have matured and the plants have dried out, *Salsola* plants begin to develop stems and flowers. They disperse their seeds in October. These plants develop flowers and mature seeds during the summer

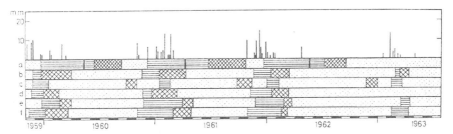

Fig. 12. Phenology of some desert plants: *top*, rainfall (mm); *a* unarmed saltwort (*Salsola inermis*); *b* hairy storksbill (*Erodium crassifolium*); *c* Hanbury's squill (*Scilla hanburyi*); *d* thread-leaved star of Bethlehem (*Ornithogalum trichophyllum*); *e* Sinai bluegrass (*Poa sinaica*); *f* short-styled sedge (*Carex pachystylis*); *hatching*, plants with leaves; *cross hatching*, plants flowering and fruiting; *stippling*, plants dormant or dead (in the case of the saltwort). *Vertical black bars* in *a* indicate the beginning of stem elongation. (After Evenari et al. 1971)

and autumn. Flowering is regulated by short days and high temperatures (Evenari and Gutterman 1976, 1985; Gutterman 1989b).

Hysteranthous geophytes, such as *Colchicum tunicatum*, flower in the hot autumn. The leaves develop during winter and seeds mature only at the beginning of the following summer (Gutterman and Boeken 1988; Gutterman 1989g). Others, such as *Urginia maritima, U. undulata* and *Scilla hanburyi* (Fig. 12c) (Evenari et al. 1971), are geophytes with bulbs that produce leaves during the rainy season, flower and produce mature seeds during the summer season (Dafni et al. 1981; Gutterman 1981; Evenari and Gutterman 1985). *U. maritima* has been found to have active roots during summer (Fridkin, pers. comm.), as has *U. undulata*.

Shrubs, such as *Zygophyllum dumosum* and *Artemisia sieberi* are perennial. They are low, and produce leaves in winter (Sect. 3.5). During summer there is a reduction in both the number of leaves and in leaf size, as well as a reduction of the total leaf area. After a winter with low amounts of precipitation, the reduction in leaf area is dramatic (Evenari et al. 1982; Orshan 1989).

Trees, such as *Tamarix* spp. inhabit wadi beds where their very deep root systems reach the water-table even during the summer. They have access to water throughout the year in their special habitats, and are consequently active throughout the year. They evaporate large amounts of water during summer (Zohary 1962).

The richness of plant species in the Negev is a result of its location where four plant phytogeographical areas meet. In addition, there are variations in topography and elevation, gradients of rain, temperature, and soil types. Together these contribute to this richness of plant species and autecological adaptations within a small area (Figs. 11, 12, 13) (Evenari et al. 1971, 1982; Shmida et al. 1986).

1.4.3 Species, Eco-genotypes and Ecotypes

The survival of any species depends on the degree of its adaptation to its habitat throughout its life cycle. Adaptation is essential in unpredictable desert conditions. One of the most important and critical stages in the life cycle of the annual plants and the renewal of perennial vegetation is by mechanisms which ensure the germination of seeds in the right micro-habitat at the proper time.

Studies have shown that only short distances may exist between populations of the same species in separate habitats in which different eco-genotypes can develop. For example, in wild barley, *Hordeum spontaneum*, in Israel: 'The results indicate that the characters studied are partly genetically determined. Striking genetic variation was found between and within populations on each site, whereas environmental interaction was found between the mesic and xeric

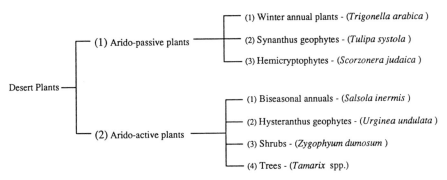

Fig. 13. Main autecology adaptations and life forms of the Negev plants (one example for each)

sites, as well as between populations and years? One of the most southerly populations of this Mediterranean plant was studied near Sede Boker. This desert population flowers, its seeds mature earlier and it produces smaller leaves than all the other populations that inhabit the Mediterranean zone (Nevo et al. 1984a). Similar results have been obtained in populations of wild wheat, *Triticum dicoccoides* (Nevo et al. 1984b). Another example is afforded by the allopatric distribution of the two varieties of *Medicago laciniata* on two different slope directions of the same hill in the Negev desert. This was investigated by Friedman and Orshan (1975) near Sede Boker.

Ecotypes and species of *Blepharis* in the Judean and the Negev deserts have been found to show different distributions. One species inhabits the slopes and has less mucilage in its seeds than others. This species disperses its seeds near the mother plant, while the other species inhabit wadi beds and develop greater quantities of mucilage on the seed coat, enabling them to be dispersed by floods (Sects. 3.3.3.4, 5.3.3) (Gutterman et al. 1967, 1969a,b).

Differences were found in the temperature and salinity suitable for germination of the seeds of two *Helianthemum* species which are well adapted to their respective habitats. *Helianthemum ventosum*, a Saharo-Arabian phytogeograhical element, mainly inhabits the south-facing slopes near Sede Boker at elevations of approximately 470–550 m above sea-level. *H. vesicarium*, on the other hand, has an Irano-Turanian phytogeographical distribution. It inhabits the north-facing slopes of hills in the same area. This is a transition zone between the Saharo-Arabian and the Irano-Turanian regions. Scarified seeds of *H. ventosum* germinate to higher percentages at high temperatures. They have higher tolerance to salinity than those of *H. vesicarium* which germinates to higher percentages in low temperatures, while germination is very low between 30°C and 35°C (Sect. 5.2.2.2) (Gutterman and Agami 1987).

Differences were also found in germination in constant temperatures and salinity as well as in alternating temperatures, in two populations of each of the two species *Helianthemum vesicarium* and *H. ventosum* in the Negev highlands of Israel, at elevations of 470 m and 920 m above sea-level. Seeds of plants from the lower elevation and more extreme desert conditions germinate faster and in higher concentrations of salinity (Sect. 5.2.2.2) (Gutterman and Edine 1988). The germination of seed populations of the Mediterranean plant *Sarcopoterium spinosum* from the Negev Desert highlands was found to be much lower (3–6%) in comparison with those from populations near Jerusalem (60%) when seeds germinated in soil (Green 1991). Mahmoud et al. (1981) found that in Saudi Arabia, *Verbesina enceliodes* may have developed ecotypes in response to altitudinal temperature gradients (Abulfatih 1983). Thurling (1966) found ecotypic differences in germination between populations of species in the genus *Cardamine*. The germination of populations from below 600 m was inhibited at high temperatures (above 25 °C), whereas seeds from populations above 1700 m did not germinate at temperatures below 10 °C (Mott and Groves 1981).

Populations of *Salvia columbariae* in the Mojave Desert and the San Gabriel Mountains of California have been compared. Each of 19 local populations observed in a study by Capon and Brecht (1970), and 10 populations observed in a later study (Capon et al. 1978), are characterized as ecotypes with individual adaptations to their specific micro-climates along the approximately 30-mile transect. The relative response to heat pretreatment of dry storage at 50 °C for up to 6 months results in high germination percentages in the desert population, but low in those of the mid-mountain.

Juhren et al. (1956) also found that, in the Californian desert, summer annuals germinate at high temperatures which are also suitable for the development of the plants, whereas winter annuals do so at low temperatures. In their detailed study, Baskin and Baskin (1971a, b, 1976, 1978, 1982) found an annual cycle of dormancy in *Cyperus inflexus* Muhl. (Cyperaceae), one of the summer annuals which grows in shallow soil 2 to 5 cm deep, in the zone bordering bare limestone in the cedar glades of middle Tennessee, North America. This plant, which germinates mainly in April, is regulated by a complex of environmental factors to reach a high percentage of germination: (1) the need for a period of stratification previous to (2) several cycles of at least 12 h of light and proper soil moisture; in addition, (3) an annual cycle of dormancy which ensures that germination occurs when there is still time for the plant to complete its life cycle and, at the same time, prevents germination in autumn and winter when plants cannot complete their life cycle. Differences in germination requirements have also been found between geographical ecotypes (and genotypes – Turesson 1992) and genetically different seasonal populations of *Chenopodium album* in Chandigarh, India (Ramakrishnan and Kapoor 1974) (Sect. 5.4.2).

High temperatures during imbibition impose thermo-inhibition in some species of the Karoo desert of South Africa which germinate in areas receiving winter rain. This ensures that, even after a temporary drop of temperature during rainy days in summer, germination of seeds of species inhabiting areas of winter rain will be inhibited. In contrast, the germination of seeds of species inhabiting areas receiving summer rain is not inhibited by high temperatures and is even accelerated (Sects. 5.4.3, 5.4.4) (Gutterman 1990a).

Seeds of different plant species have been found to require different conditions for germination. These enable them to germinate in the right period of the right season, and in the right place, and to survive in extreme hot deserts with completely unpredictable dates and distribution of rainfall (Sect. 5.2). Populations of the same species were found to adapt according to the slope direction, elevation, etc. It is, therefore, interesting to compare the germination mechanisms of plant species inhabiting different localities along the northern part of the Saharo-Arabian desert, which receives about 100 mm of rain in winter, with those inhabiting the southern part of the Sahara and the Saudi-Arabian Peninsula, which receive about the same amount of rain in summer. This is similar to the amount (about 100 mm) received in Sede Boker and most of the Negev Desert highlands. One can divide the hot deserts of the world into four groups, according to the season of rainfall: (1) Desert areas receiving winter rains; (2) Desert areas receiving summer rain; (3) Desert areas receiving infrequent rain; (4) Deserts receiving rain throughout the year (Berkofsky 1983; Abd el Rahman 1986; Gupta 1986; le Houérou 1986; Monod 1986; Orshan 1986; Walter 1986; Werger 1986).

1.4.4 Genotypic and Phenotypic Influences on Seed Germination

In addition to the genotypic influences mentioned above, which increase the chances of a seedling to survive by germinating at the right time an the right place (Sects. 4.1.3, 4.2, 5.1.2, 5.1.3), phenotypic, maternal and environmental factors have an influence that leads to the spread over time of the germination of the seed population of the species (see Chap. 2). The result is that only a portion of the seed population will germinate after one rain event. The same phenomenon is also achieved in other plant species by mechanisms of seed dispersal by rain. In this case, only a few seeds are released and germinate after a single shower (Sects. 3.2.6, 3.3.2.3, 3.3.3.4). In yet other species, germination is regulated by germination inhibitors, periods of after-ripening, or physical barriers of the seed coat, such as impermeability to water or O_2 (Sects. 2.3.1, 2.4.2, 4.2). These regulator mechanisms are all very important for survival, especially under extreme and unpredictable desert conditions (see Chap. 6). The importance of spreading germination over time, and of reducing the risk of mass germination, is obvious.

The inheritance, as well as the history of each seed from the beginning of its development, the conditions of maturation, storage, and imbibition, all affect 'the seed's readiness for germination' at, and within, a given time.

2 Phenotypic Effects on Seeds During Development

2.1 Introduction

If genotypic inheritance increases the fitness of plant species to their natural habitats so that germination occurs in the right place and at the right time (Sect. 5.1.2), phenotypic effects should, in addition, increase the diversity of seed germination. In many species the fate of the following generation or generations, as far as the germination of seeds is concerned, is dependent, at least to some degree, on the position of the seeds, as well as on their conditions of maturation while still on the mother plant. These conditions include environmental factors such as day length, temperature, the position of the seeds in the fruit and inflorescence and in the position of the fruit or inflorescence on the mother plant.

The germinability of seeds has been found to be affected by: (1) the position of the inflorescence on the mother plant (Thomas et al. 1979; Jacobsohn and Globerson 1980; Grey and Thomas 1982; Gutterman 1990a), (2) the position of the flowers and seed in the inflorescence (Evenari 1963; Koller and Roth 1964; Datta et al. 1970; Evenari et al. 1977) or even the position of seeds in the fruit (Gutterman 1980/81a), (3) the age of the mother plant during flower induction (Kigel et al. 1979) and (4) the age of the plant during the last stage of seed maturation (Gutterman and Evenari 1972; Do Cao et al. 1978; Gutterman 1978a). The order and position of the caryopsis from which the mother plant originated may also influence seed germinability (Datta et al. 1972a).

Seed germinability has also been found to be affected, during the development and maturation, by environmental factors including: (1) day length (Lona 1947; Jacques 1957, 1968; Koller 1962; Cumming 1963; Wentland 1965; Evenari et al. 1966; Gutterman 1969, 1973, 1974, 1978a,b,c, 1982a, 1985; Karssen 1970; Gutterman and Evenari 1972; Gutterman and Porath 1975; Pourrat and Jacques 1975), (2) temperature (Juntila 1973; Heide et al. 1976), (3) parental photothermal environment (Datta et al. 1972a; Wurzburger and Koller 1976; Kigel et al. 1977), (4) light quality (Cumming 1963; McCullough and Shropshire 1970; Gutterman 1974; Gutterman and Porath 1975; Jacobsohn and Globerson 1980; Cresswell and Grime 1981) and (5) altitude (Dorne 1981). Achenes of *Lactuca serriola*, which mature during summer and autumn (Gutterman 1992c, in press), as well as the summer and winter maturing seeds (on the same mother plants) of certain biseasonal flowering perennial shrubs (Aizoaceae) have been found to show different germinability (Gutterman 1991).

26   Phenotypic Effects on Seeds During Development

The maturation of seeds having different germinability on one mother plant has a very important ecological advantage, especially in deserts (Gutterman 1980/81 b, 1982a, 1983, 1985, 1986a, b; Roach and Wulff 1987; Fenner 1991) where the date of the first rains of the season that causes germination, the amount of rain and its distribution are unpredictable. In the Negev Desert highlands, the onset of rain which causes germination may range from mid-November to the end of February (3.5 months). In one season, only one rain-fall may cause germination. In another, two to even three or more rain events may cause germination (Gutterman 1982b and Tables 1–5) (Chap. 1). An almost identical phenomenon has been found in species of weeds which survive in unpredictable conditions (Gutterman 1985). Different maternal and environmental factors have been found to affect the germination of seeds in different ways. This ensures that, even under the most suitable conditions, only a portion of the total seed bank of the population of seeds of a certain plant species will be ready to germinate at one time. It is not only the position and conditions of maturation that influence seed germinability: seeds which develop on a plant from female or hermaphrodite flowers have been found to show differences in dispersal, germinability and longevity (Roiz 1989).

2.2 Position Effects

The position of seeds on the mother plant may affect their size, morphology and germination. The position of the inflorescence and fruit, as well as the position of the seed in the inflorescence or fruit have been found to influence seed germinability.

2.2.1 Position Effects in Plants with Aerial and Subterranian Fruits (Amphicarpy)

In heterocarpic species, which are characteristic of semi-arid and arid floras, each plant produces two types of diaspores. One is telechorous for long-distance dispersal, and the other topochorous. By this, long-distance dispersal is prevented. In still others, limited dispersal is achieved in a variety of ways (Zohary 1937, 1962). In amphicarpic species, the subterranean topochorous diaspores germinate in situ, whereas the aerials are telechorous.

Gymnarrhena micrantha Desf. (Compositae), is an annual which inhabits the Negev Desert highlands and has a Saharo-Arabian distribution extending into W. Irano-Turanian areas (Feinbrun-Dothan 1978). In this amphicarpic species, the telechoric aerial achenes (seeds) are dispersed by wind after being released from the closed capitulum by rain. They differ from subterranean atelechoric achenes which germinate in situ from the dead mother plant (Zohari 1937). Aerial achenes with a well-developed pappus are much smaller (0.37 mg) than subterranean achenes (6.50 mg) which have an undeveloped pappus (Fig. 14).

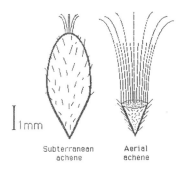

Fig. 14. *Gymnarrhena micrantha.* Comparison of aerial and subterranean achenes: a schematic drawing

In a range of temperatures from 5 to 25 °C; in light, the greatest difference was observed between the final germination of the subterranean achenes at 25 °C (87%) and of the aerial achenes (38%). In the dark, on the other hand, 30% subterranean achenes germinated, but only 4% aerials. At lower temperatures, the germination of both types of achenes was higher in the dark and reached between 75 and 82%. Germination was above 90% at 10 °C in light. The lower the temperature, the closer was the percentage of germination of aerial and subterranean achenes. At 5 °C, almost all achenes of both types germinated at the same time (Koller and Roth 1964).

Emex spinosa, (L). Campd. (Polygonaceae), has a Mediterranean distribution extending into the Saharo-Arabian region (Zohary 1966). It is an amphicarpic species. The subterranean propagules germinate in situ from the dead mother plant. They are smooth and much larger (75 mg) than the arials. The latter are spiny, and range in weight from 24 mg (lower propagules) to 2 mg (terminal arials). The aerial propagules are dispersed by wind, flood water or animals. Their germinability is much higher in all conditions tested than that of the subterranean propagules in populations which inhabit the Negev Desert (Table 6) (Evenari et al. 1977). The substances leached from the aerial propagules contain germination inhibitors, but not the substances leached from subterranean propagules. Germination is much lower at a constant temperature than in alternating temperatures in both aerial and subterranean prop-

Table 6. *Emex spinosa* germination (% ± S.E.) of non-leached aerial or subterranean propagules after 7 days of imbibition under alternating or constant temperatures; imbibition in the light or dark. (After Evenari et al. 1977)

First temperature (°C) for 18 h	Second temperature (°C) for 8 h	Aerials light	Aerials dark	Subterraneans dark
15	30	34 ± 1.0	60 ± 2.1	20 ± 1.4
15	15	6 ± 2.0	10 ± 3.2	0
20	35	35 ± 2.0	54 ± 1.2	26 ± 1.5
20	20	8 ± 2.0	24 ± 2.0	7 ± 1.8

Table 7. Germination (% ± S.E.) of aerial and subterranean propagules of *Emex spinosa* after 7 days of imbibition leached, unleached, or unleached with micropylar tip cut open; at various temperatures (5 to 35 °C) in light (*L*) and dark (*D*). (After Evenari et al. 1977)

Temp. (°C)	Germination %							
	Aerial seeds						Subterranean seeds	
	Unleached		Leached		Unleached with micropyler tip cut open		Unleached with micropyler tip cut open	
	L	D	L	D	L	D	L	D
5	0	0	0	0	78 ± 2	75 ± 3	18 ± 2	68 ± 1
10	0	0	5 ± 0.8	13 ± 1.1	80 ± 3	78 ± 3	24 ± 1	74 ± 3
15	2.0 ± 1.0	5 ± 1	20 ± 1.7	44 ± 1.8	84 ± 2	82 ± 3	40 ± 2	80 ± 4
20	7.5 ± 1.5	24 ± 1.5	36 ± 1.5	60 ± 2.0	85 ± 1	85 ± 1	80 ± 1	85 ± 3
25	7.5 ± 2.0	3 ± 1.5	36 ± 2.0	41 ± 1.2	85 ± 2	85 ± 3	80 ± 4	80 ± 1
30	0	0	30 ± 1.9	17 ± 1.9	80 ± 3	80 ± 4	70 ± 2	80 ± 3
35	0	0	10 ± 2.0	5 ± 0.7	–	–	–	–

agules. Germination in white light is lower than in the dark at lower temperatures and higher at high temperatures (Table 7) (Evenari et al. 1977). When aerial propagules were leached away, germination increased. When the micropylar tip of the aerials was cut open, inhibition by white light disappeared. In the subterranean propagules, light inhibition increased as the temperature was lowered from 20 °C to 5 °C (Table 7) (Evenari et al. 1977). The lower germination of the subterranean atelechoric propagules, and their low numbers per plant are important in preventing competition. They ensure a wider period of germination.

The inhibitors that occur in aerial telechoric propagules have an influence on the amount of rain or washing by floods that is needed before these propagules germinate. These germination inhibitors may act as a 'rain gauge' or 'rain clock' which ensures that germination will take place only after sufficient wetting for the establishment of the seedlings. In addition to better germination in the dark and the long period (7 days) of imbibition required for germination, this provides a better chance for buried propagules to germinate in favourable microhabitats, such as depressions.

2.2.2 Position of the Capsules on the Plant Canopy Affects Germination

Glottiphyllum linguiforme (Aizoaceae), is a perennial shrub inhabiting areas of South Africa. On each plant there are **central** and **peripheral** capsules. These differ in size, number of loculi valves and number of seeds, as well as in seed germinability (Figs. 15, 16).

Very low percentages of seeds taken from central capsules germinated in petri dishes at 25 °C after 18 days, in both light and dark. In the same period, nearly 80% of seeds from the peripheral capsules germinated in the light, and

Fig. 15. *Glottiphyllum linguiforme* capsules. *Left* wet and open; *right* dry and closed. The *upper pair* are the peripheral large capsules which are easily detached from the shrub as a dispersal unit. The *lower pair* are the small capsules which, with time, are situated beneath the shrub and are possibly the seed bank which could replace the mother shrub after its death (×2)

70% in the dark with short illuminations (Fig. 16) (Gutterman 1990a). Seeds from central capsules that had matured during the previous 3 years of the experiment did not germinate when placed in wet soil. However, approximately 20% germination took place when the peripheral capsules adhered to wet soil.

The capsules in the peripheral part of the plant are larger and contain approximately 200 seeds, whereas those from the central part of the shrub are smaller and contain approximately 125 seeds. The loculi valves of both capsules opened 6–7 min after wetting. The seeds did not disperse, however, since they are covered by two loculi-roofs and blocked by placental tubercles. The peripheral capsules are easily separated from the mother plant, and it is possible that they may act as a unit which is dispersed by wind or floods. The central dispersal units remain below the canopy in the middle of the shrub. They probably form a local seed bank which supplies seedlings to replace the dead mother plant (Gutterman 1990a).

2.2.3 Position Effects of Seeds in the Inflorescence When It Is a Dispersal Unit

Pteranthus dichotomus Forssk. (Caryophyllaceae) is a Sahoro-Arabian annual which extends into some Irano-Turanian, Mediterranean and Sudanian ter-

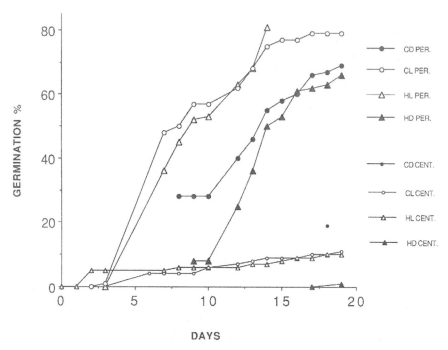

Fig. 16. Influence of high 45 °C (*H*) or low 25 °C (*C*) temperature and light (*L*) or dark (*D*) on the germination of seeds of *Glottiphyllum linguiforme* from peripheral capsules (*PER*) and central capsules (*CENT*). (After Gutterman 1990a)

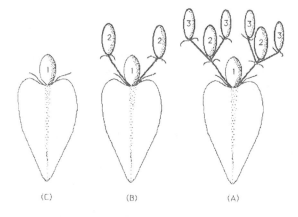

Fig. 17. *Pteranthus dichotomus.* Winged thorny spike of a dispersal unit (*A*) showing the arrangement of the pseudocarps containing one seeded fruit each, in the order *1, 2* and *3*. In dispersal units (*C*) only order *1* develops and in (*B*), orders *1* and *2*. (After Evenari et al. 1982)

ritories (Zohary 1966). Under favourable conditions, one dispersal unit (an inflorescence) contains seven pseudocarps, each containing a single seeded fruit arranged in three orders (Fig. 17A). The first order contains one pseudocarp, the second order two pseudocarps, and the third order four pseudocarps. These are the terminals. In unfavourable conditions, only one or two orders

Position Effects

Table 8. Germination % in temperatures from 8 to 37°C in Light (*L*) and Dark (*D*) of *Pteranthus dichotomus* pseudocarps containing one seeded fruit of the orders 1, 2 and 3 in a dispersal unit with three orders (see Fig. 17). (After Evenari et al. 1982)

Temp. (°C)	Order of pseudocarps and germination (%)					
	1		2		3	
	D	L	D	L	D	L
8	12	18	58	72	99	90
15	23	50	83	89	97	98
26	8	65	90	97	95	100
30	16	72	84	97	96	100
35	4	12	20	66	89	100
37	0	0	24	65	82	91

develop (Fig. 17B, C). The terminals of each dispersal unit always germinate better than the subterminals. In a dispersal unit having all three orders, the terminal 'seeds' (order 3) achieved a higher percentage of germination, both in the light and dark, in a range of temperatures from 8 to 37°C, compared with the germination of the subterminals (orders 2 to 1). The poorest germination was observed in 'seeds' of order 1. The higher the order, the greater the percentage of germination in light or dark, at all temperatures (Table 8). In dispersal units of one 'seed' (order 1) the 'seeds' germinate as well as those in the terminals of order 3 (Evenari 1963; Evenari et al. 1982). This positional effect ensures that only one or two seedlings will appear from seven 'seeds' in a single dispersal unit during one season. This fact has already been observed in natural habitats (Evenari et al. 1982).

Aegilops geniculata Roth (= *A. ovata* L.) (Poaceae), is a Mediterranean annual which extends into the West Judean Desert (Feinbrun-Dothan 1986). The spike, which is the dispersal unit, is composed of 2−4 spikelets. Each of the lower spikelets of the 3−4 spikelet dispersal units contains two caryopses (grains), a_1, a_2 and b_1, b_2. The terminal spikelet contains only one caryopsis, b, c, and d (Fig. 18). When isolated spikelets were grown in soil (in pots) under natural day length (ND) (between 10 h and 12.5 h), it was found that seedling emergence was highest (88% ±4) from spikelet-(A) and lowest (10% ±6) from spikelet-(C). The time of emergence of the coleoptile and other stages of development, such as age at anthesis, are faster in plants originating from spikelet-(A) than in those originating from (B). Those from (C) are the slowest (Table 9) (Datta et al. 1972a; Gutterman 1992a).

When caryopses were sown under short days (8 h) or long days (20 h), a_1 and b_1 caryopses gave higher percentage of seedling emergence and flowered more quickly than did caryopses a_2 and b_2. Plants originating from caryopsis

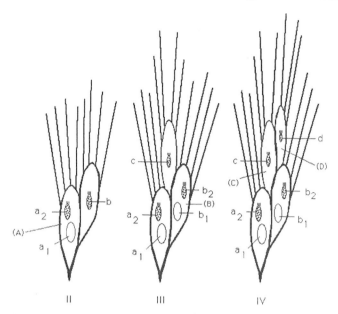

Fig. 18. Schematic drawing of the three types of spikes (*II, III, IV*) of *Aegilops geniculata* (*A. ovata*). Position and number of caryopses ($a_1 - d$) is shown in the spikes containing *II, III* or *IV* spikelets. (After Datta et al. 1970; Gutterman 1992a)

Table 9. Emergence (% ± S.E.) of *Aegilops geniculata* seedlings under natural day length (ND) 10 – 12.5 h from the isolated spikelets of a 3-spikelet dispersal unit. Time interval (days ± S.E.) between sowing and emergence of coleoptile and the age of plant at time of flowering. Temperature during experiment 20 – 30 °C (see Fig. 18). (After Datta et al. 1972a)

Type of spikelets	Emergence	Emergence time	Age at anthesis
A	88 ± 4	3.1 ± 0.1	83.1 ± 1.6
B	72 ± 7	3.7 ± 0.2	94.0 ± 1.8
C	10 ± 6	4.4 ± 0.2	109.0 ± 2.7

c showed the lowest percentage of seedling emergence after the longest time. Moreover, flowers appeared after the longest time. Under long and short days, plants originating from b_1 caryopses flowered first (Table 10) (Datta et al. 1972b).

When a_1 and b_1 caryopses were soaked in water at constant temperatures from 5 to 35 °C, both in the light and in the dark, a higher percentage of a_1 germinated than b_1 after 8 days (Table 11), with but two exceptions.

When caryopses a_1 and a_2 from a dispersal unit with two spikelets were compared with caryopses a_1, a_2, b_1 and b_2 from a dispersal unit of three spikelets, after 24 h of imbibition at temperatures of 15 to 35 °C the lower

Position Effects

Table 10. Effect of 20 h-LDs and 8 h-SDs on seedling emergence (% ± S.E.) of *Aegelips geniculata* from various types of caryopses, the time until emergence (days) and age (days) at flowering, of plants derived therefrom (see Fig. 18). (After Datta et al. 1972a)

Type of caryopses	Seedling emergence (%)		Time (days) of coleoptile emergence		Age (days) of flowering	
	LD	SD	LD	SD	LD	SD
a_1	96 ± 4	88 ± 17	3.1 ± 0.1	2.6 ± 0.1	33.2 ± 1.4	95.2 ± 4.3
a_2	79 ± 5	56 ± 20	4.1 ± 0.2	3.8 ± 0.2	35.1 ± 1.7	99.9 ± 5.4
b_1	91 ± 7	96 ± 2	3.2 ± 0.1	3.2 ± 0.1	32.1 ± 2.0	89.7 ± 6.8
b_2	80 ± 0	52 ± 8	4.1 ± 0.1	3.7 ± 0.1	38.9 ± 2.5	97.3 ± 7.8
c	57 ± 16	34 ± 12	4.3 ± 0.2	4.4 ± 0.3	41.3 ± 3.6	120.4 ± 10.9

Table 11. Germination of caryopses from dispersal units of *Aegilops geniculata* after 8 days of imbibition (see Fig. 18). (After Datta et al. 1970)

Temperatures (°C)	Light		Dark	
	a_1	b_1	a_1	b_1
5	0	0	0	0
10	75	70	90	65
15	75	35	70	75
20	55	30	70	40
25	10	5	65	50
30	0	5	30	15
35	0	0	0	0

caryopses a_1 and b_1, showed a higher percentage of germination than caryopses a_2 or b_2 (Table 12) (Datta et al. 1970).

When lettuce (*Lactuca sativa* L.) achenes imbibed the solution leached from hulls of spikelets, a, b and c of a three-spikelet dispersal unit (25.6 g hulls per 300 ml water) in the light, at a temperature of 26 °C, an inhibitory effect was observed compared with the water control. The inhibitory effect is the highest in hulls from spikelet (c) and lowest from those of spikelet (a) (Table 13) (Datta et al. 1970).

L. sativa achenes were allowed to imbibe the solution leached from the hulls of spikelets III a and III b (25.65 g hulls per 100 ml water) in the light and at 20 °C. Again, in the two concentrations measured, an inhibitory effect was observed. This was due to chemicals which had entered the water from the hulls. The inhibiton was higher from hulls from spikelet III b in comparison with III a (Table 14) (Datta et al. 1970). Inhibition is much less in dark.

The main inhibitor was found to be mono-epoxylignanolid (MEL). This inhibits *Lactuca sativa* cv. 'Great Lakes' in incandescent light but not in darkness. In green, but not in blue or red light, there is an inhibitory effect

Phenotypic Effects on Seeds During Development

Table 12. Effect of various temperatures on the germination (%), after imbibition for 24 h of caryopses of *Aegilops geniculata* from a dispersal unit of two (II) or three spikelets (III) (see Fig. 18). (After Datta et al. 1970)

Dispersal unit type	Type of caryopses	Temperatures and germination (%)			
		5°C	15°C	25°C	35°C
II	a_1	0	100	96	94
	a_2	0	64	84	50
III	a_1	0	96	92	80
	a_2	0	16	60	24
	b_1	0	94	98	84
	b_2	0	24	86	60

Table 13. Effect of leachate from the hulls of three-spikelet dispersal unit of spikelets *a*, *b* and *c* (25.65 g hulls per 300 ml water) on the germination (% ± S.E.) of *Lactuca sativa* achenes in light at 26°C after 24 h (see Fig. 18). (After Datta et al. 1970)

Leachate of hull used	Germination (%)
a	48 ± 3
b	36 ± 4
c	29 ± 2
Control (water)	94 ± 2

Table 14. Effect of half and quarter dilutions of leachate of 25.65 g per 100 ml water, from hulls of spikelets *a* and *b* of three-spikelet dispersal units, on the germination (% ± S.E.) of *Lactuca sativa* at 20°C in light after 24 h (see Fig. 18). (After Datta et al. 1970)

Leachate dilutions	Hulls	
	Leachate from a	Leachate from b
1/2	63 ± 2	29 ± 4
1/4	84 ± 4	70 ± 4
Control (water)	91 ± 1	91 ± 1

which lasts for 24 h in far red (FR) light. This is followed by 24 h of imbibition in red light (Lavie et al. 1974; Gutterman et al. 1980). The higher the osmotic potential of NaCl in the water which the caryoposes imbibed, the bigger the difference in the germination (after 72 h) between caryopses a_1 and a_2, both in light and darkness (Table 15).

Position Effects

Table 15. Effect of osmotic pressure on the germination ($\% \pm$ S.E.) after 72 h of a_1 and a_2 grains of three-spikelet dispersal units of *Aegilops geniculata* at 20 °C (see Fig. 18). (After Datta et al. 1970)

Osmotic pressure of NaCl solutions (atm.)	a_1		a_2	
	L	D	L	D
0	97 ± 3	98 ± 2	78 ± 3	82 ± 3
5	94 ± 1	99 ± 1	52 ± 7	55 ± 7
10	93 ± 2	90 ± 0	34 ± 5	29 ± 7
15	68 ± 4	90 ± 2	3 ± 1	15 ± 4

Influences of One Generation on the Other. A. geniculata (*A. ovata*) caryopses a_1, b_2 and c of three-spikelet dispersal units were compared according to their order in the mother plant. The position effect of the caryopses in a dispersal unit not only had a direct influence on the germinability of the caryopses and on the stages of development, but also an influence on the germinability of the caryopses of the second generation. This depended on the caryopsis from which the mother plant had originated, and the conditions in which they were grown.

Large differences were found in germinability under temperatures similar to those existing during the growing season in the natural habitat (15/10 °C). These were not found when plants were grown at higher temperatures (28/22 °C). The origin of the mother plant was also found to have an influence

Table 16. Effect of position on average weight (mg $\pm$ S.E.) and germination (%) after 24 h in light at 15 °C of *Aegilops geniculata* ($= A. ovata$) caryopses harvested from plants originating from a_1, b_2 and c caryopses and grown under 18 h LDs at temperatures of 15 °C during the day and 10 °C during the night, and at temperatures of 28 °C during the day and 22 °C during the night. All caryopses mentioned are separated from three-spikelet dispersal units (see Fig. 18). (After from Datta et al. 1972a)

Order of caryopses from which mother plant developed	Order of caryopses collected from mother plant	Average weight of caryopses (mg)		Germination (%)	
		15/10 °C	28/22 °C	15/10 °C	28/22 °C
a_1	a_1	20.6 ± 0.7	13.9 ± 0.5	84.4	100.0
	b_2	9.5 ± 0.8	6.7 ± 0.3	10.0	60.0
	c	6.1 ± 0.6	3.0 ± 0.3	8.5	63.1
b_2	a_1	22.9 ± 0.5	12.9 ± 0.4	55.0	100.0
	b_2	9.2 ± 0.2	7.1 ± 0.2	2.3	85.3
	c	3.8 ± 0.4	3.0 ± 0.2	0	90.0
c	a_1	27.3 ± 1.9	14.7 ± 1.7	21.2	100.0
	b_2	13.0 ± 0.3	6.9 ± 0.8	0	76.0
	c	4.2 ± 0.7	3.5 ± 4.2	0	86.7

on the weight of caryopses a_1, b_1 and c. In this case the position effect, together with the environmental factors during growth and caryopsis maturation, has influences on future generations (Table 16) (Datta et al. 1972a).

The different caryopses of a dispersal unit differ in colour, size and weight according to their position, the conditions in which maturation took place, and the caryopsis from which the mother plant originated. Germination is also affected by the position and the origin of the different caryopses of the dispersal unit, as well as environmental factors present during their maturation. There are also differences in the inhibitory effects of the hulls of the different spikelets of the dispersal unit. The higher the osmotic pressure, the greater the differences in germination between caryopses a_1 and a_2, both in light and in the dark. All of these components are involved in the heteroblasty of the caryopses of the plant. They ensure that the germination of the caryopses of each spike is dispersed in time. Simultaneous germination of all the caryopses of the dispersal unit is thus prevented. Only one or two seedlings from $5-6$ caryopses were observed in the field in a single season.

2.2.4 Position of Achenes in the Capitulum Whorls and Their Germination

In *Asteriscus pygmaeus* there are mechanisms that delay achene dispersal and spread dispersal and germination over time (Sect. 3.3.3.2). The capitula are closed when dry and open when wet. During some rain events, a few of the peripheral achenes are dispersed (Fahn 1947). Only disconnected achenes germinate (Koller and Negbi 1966). The percentage of germination of achenes from the peripheral whorl is much higher than of achenes from the subperipheral whorl (Fig. 91). Each year, some of the achenes are disconnected and dispersed by rain (Fig. 83). The seed bank of this desert annual can remain protected in the capitula of the dead mother plant for many years (Figs. 84, 85, 86) (Gutterman and Ginott in press).

2.2.5 Position of Female and Hermaphroditic Flowers and Their Seed Germinability

Parietaria judaica Mert. et Koch, is a perennial herb found in shady habitats. It occurs in the Mediterranean and Irano-Turanian, extending into the Sudanian and Euro-Siberian regions. This species is also found in the Judean desert. It is gynomonoecious. On the same plant, and on the same inflorescence, there are female flowers (F) and hermaphroditic flowers (H). The central female flower opens first and the hermaphroditic flowers $2-4$ days later. This species is used in folk medicine for healing wounds (Zohary 1966). Seeds from female flowers were found to have higher germinability than seeds from the hermaphroditic flowers (Fig. 19) (Roiz 1989). The same is true of their longevity; after maturation or 1 or 2 years of storage, the seeds from female flowers germinate earlier than those from hermaphroditic flowers

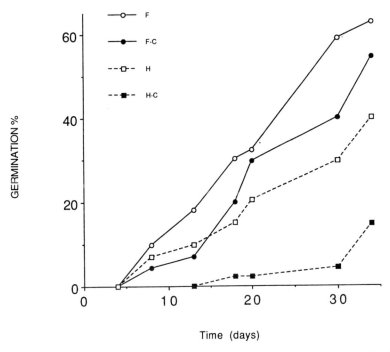

Fig. 19. Germination of *Parietaria judaica* seeds originating from hermaphroditic flowers with covers ($H \cdot C$) or without covers (H), in comparison with seeds originating from female flowers with covers ($F \cdot C$) or without covers (F). Germination in petri dishes at 25 °C, 12 h light and 12 h dark. (After Roiz 1989)

(Fig. 20) (Roiz 1989). The seeds from female flowers are mainly heterozygotic. They are more resistant in unpredictable conditions than seeds from hermaphroditic flowers. Furthermore, the dispersal units of the seeds originating from the female flowers are more hairy and can be dispersed further. This correlates with the fact that the seedlings do not compete well with the adult plants. Seeds from female flowers produce plants that grow well, far away from the adult plants, and are resistant to water stress. This is not the case with the seeds from the hermaphroditic flowers. These have less hairs on their dispersal units. Mainly, they disperse only over short distances and their seedlings grow well near the adult plant (Roiz 1989).

2.2.6 Position of Flowers and Dimorphism

Salicornia europaea (Chenopodiaceae) is a highly salt-tolerant annual halophyte, and one of the pioneers that occupy dried-up saline marshes. This plant also inhabits marshes in desert areas, such as Central Negev, Lower Jordan Valley and the Dead Sea. It is a Mediterranean and Euro-Siberian plant

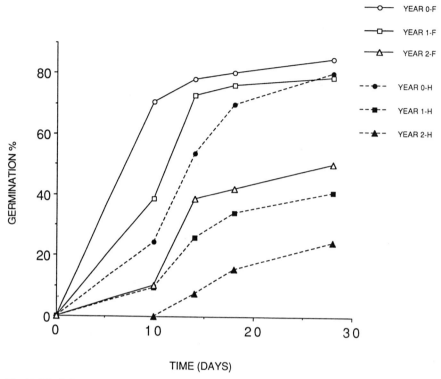

Fig. 20. The influence of storage periods on the germination of *Parietaria judaica* seeds developed from female flowers (*F*) and hermaphroditic flowers (*H*). Seeds that germinated in pods in greenhouse conditions immediately after maturation (year 0) or after 1 or 2 years of storage. (After Roiz 1989)

that has been introduced elsewhere. It flowers in groups of three, the middle flower situated above two laterals (Zohary 1966). The single seed of the median flower is large (mean of 1.8±0.1 mm) and weighs a mean of 0.78±0.1 mg. The single seed of the lateral flowers is small (mean of 1.1±0.1 mm) and weighs a mean of 0.24±0.04 mg.

The large seeds germinate to about 90% after 59 days of soaking in distilled water and the small seeds germinate to only 50% in the same time. This was observed in plants collected in the North American Morton Salt Company, Rittman, Ohio. It was also found that after 6 weeks of stratification and 1 week of imbibition, in light, large seeds germinate to 74% and in the dark to 53%. The small seeds germinate to 30% in light and only 16% in the dark. The recovery of seeds after 56 days in 5% NaCl, when soaked in distilled water for 42 days, was 91% of germination for the large seeds and only 16% for small seeds. The small seeds appear to be much less salt tolerant than the large ones. The large seeds germinate to higher percentages in NaCl concentrations such as 1, 3, 5% than the small ones. (Ungar 1979; Philipupillai and Ungar 1984).

Position Effects

Atriplex dimorphostegia (Chenopodiaceae) is an annual desert plant of the sandy and/or saline areas of the Negev and Arava valley. It derives from the Irano-Turanian and Sahara-Arabian plant geographical regions (Zohary 1966). Two types of dispersal units are formed on a single plant: flat or humped. The humped type appears on the distal ends of the branches and the flat type below them. The flat type appears and matures earlier. When the fruit is separated from the dispersal unit, the one seed of the flat type germinates to 20% at 20°C in light, compared with only 6% for the humped type. In the dark, 68% and 38% respectively germinate (Koller 1954, 1957; Koller and Negbi 1966) (Sect. 5.2.2.1).

Dimorphic dispersal units are also formed in *Atriplex rosea* (Kadman 1954), *A. semibaccata*, *A. holoscarpa*, as well as in *A. inflata* in Australia (Beadle 1952), in which the fruits also differ.

2.2.7 Position, Heteromorphism and Germinability

The winter annual composite *Hedypnois cretica*, which inhabits Mediterranean and desert areas of Israel, produces three different diaspore-morphs: (1) the smallest (1.03 mg) inner achenes that have a pappus and the highest percentage of germination (77–86%) at 15°C in light; (2) the larger (2.14 mg) outer achenes with the lowest germination (41–46%); and (3) the largest (4.48 mg) marginal epappose achenes with germination of 42–51%. There are no significant differences in the percentages of germination at 15°C in light between each of the three types of achenes in plant populations along gradients of increasing aridity: Jerusalem (540 mm rain), Yatir (272 mm) and Arad (144 mm) (Kigel 1992).

In the Namaqualand desert of South Africa, the three types of diaspores of *Dimorphotheca polyptera* also enable the species to spread germination in time and location (Beneke 1991).

2.2.8 Position Effect of Seeds in the Fruit

Mesembryanthemum nodiflorum L. (Aizoaceae) is an annual desert plant whose seeds mature in closed capsules at the beginning of summer. This plant species originates from South Africa and inhabits areas in the Mediterranean, W. Europe-Siberian (Atlantic) and Saharo-Arabian geographical regions (Zohary 1966). After dispersal the seeds adhere to the soil crust until they germinate many years later (Sect. 3.3.3.1). The capsules are closed when dry, and open when wet. The seeds in the capsules are protected from the time of maturation until the following winter when they are dispersed by rain. Dry capsules (fruit) were harvested from natural populations in the Judean Desert near Jericho. In a single fruit there are 30~ seeds, 10~ of each group. Much higher percentages of the **terminal seeds** germinate than of the seeds from the middle and lower groups. Under laboratory conditions, the terminal seeds are dispers-

40   Phenotypic Effects on Seeds During Development

Table 17. Effect of position in the capsules of *Mesembryanthemum nodiflorum* seeds on the time of wetting needed for upper (*I*), central (*II*) and lower (*III*) groups of seeds to be released from the capsules. The germination (% ± S.E.) of seeds of each group after 3 – 12 days of imbibition at 25 °C in the light: (*A*) Experiment carried out on 25 December 1979. (*B*) Experiment carried out on 14 January 1980. (After Gutterman 1980/81 a)

Seed position and time of release (min)		Germination (%) after (days)			
		A		B	
Seed position	Release time (min)	8	12	3	8
(I)	15	29.0	61.0	8.0	57.5 ± 4.0
(II)	200	3.2	5.5	1.5	9.0 ± 1.3
(III)	320	1.0	1.0	0.5	8.7 ± 2.4

ed first, after 15 min of wetting, the seeds from the middle part of the fruit are dispersed after about 200 min of wetting, and the lowest group of seeds only after 320 min.

Eight years after maturation and storage in the dry capsules, the seeds were separated and allowed to imbibe water in petri dishes. Much higher percentages (61%) of the terminal seeds germinated than of seeds from the middle (5.5%) or lower (1%) parts of the fruit (Table 17) (Gutterman 1980/81 a). With time, the percentages of germination in the other two groups of seeds increased. Each of these three groups of seeds has a different period of after-ripening (Sect. 5.6) (Gutterman 1990a, b).

2.2.9 Position and Annual Rhythm in Seed Germinability

Even after 16 years of storage under laboratory conditions, germination of seeds of the terminal group is rapid and high in winter compared with summer

Table 18. Germination (% ± S.E.) of terminal seeds of *Mesembryanthemum nodiflorum* after 9 days of imbibition under continuous light at constant (*A*) or alternating (*B*) temperatures. (After Gutterman 1980/81 a)

Month of germination	A			B	
	Germination at constant temperatures			Germination at alternating temperatures	
	15 °C	25 °C	35 °C	15 → 35 °C	35 → 15 °C
April	77.0 ± 2.4	69.0 ± 8.8	89.0 ± 6.0	58.0 ± 5.8	99.0 ± 2.4
June	11.5 ± 1.0	16.5 ± 2.8	6.5 ± 0.5	19.0 ± 4.7	30.5 ± 5.7
September	12.0 ± 1.7	14.5 ± 2.6	5.5 ± 1.8	13.5 ± 3.1	16.0 ± 3.9
December	66.5 ± 12.0	81.0 ± 3.9	–	64.0 ± 15.6	84.5 ± 5.0

Environmental Effects 41

(Table 18). This has not been observed in the two other groups of seeds of the same fruit (Gutterman 1980/81 a, 1990 a, b).

2.3 Age Effects

2.3.1 Senescence of the Mother Plant Affecting Seed Germination

Seeds of *Trigonella arabica* (original data) and *Ononis sicula* (Evenari et al. 1966; Gutterman 1966, 1980/81 a) that mature under LD when the plant has started to dry out at the end of the season, have incomplete seed coats, which is typical of maturation under SD in younger plants. Imbibition as well as germination of these seeds is much faster than in the typical yellow seeds that have matured under LD (see also Sect. 2.4.3 (Table 24) (Gutterman 1966).

2.3.2 Age of the Mother Plants Affecting Seed Germination Was Also Found in Non-Desert Plants

Subterranean clover (*Trifolium subterraneum*) seeds that developed on the mother plant at an early stage contained a higher proportion of hard seeds than those seeds that matured later (Aitken 1939). The same was also observed in milk vetch (*Astragalus sinicus*).

Amaranthus retroflexus L. (Amaranthaceae), origin N. America, but now pluriregional, is one of the dominant weeds of irrigated summer crops (Zohary 1966). The later the flower induction by 1–3 SD at the age of 6–15 days, the lower the seed germination (Kigel et al. 1979).

Oldenlandia corymbosa (Rubiaceae) is from the tropical zones of Africa, America and Asia. Less dormant seeds develop in older plants in comparison with younger plants from the line that produces dormant seeds (Do Cao et al. 1978).

2.4 Environmental Effects

Seed germination has been found to be influenced by environmental conditions on the mother plant during its growth as well as during maturation of the seeds.

2.4.1 Photothermal and Position Effects

In *Aegilops geniculata* (*A. ovata*), as mentioned in Sect. 2.2.3, day length, temperature and position have a large influence on the germinability of the caryopses at low temperatures (15/10 °C) but not at 28/22 °C, (Fig. 18, Table 16) (Datta et al. 1970, 1972a; Gutterman 1992a).

42 Phenotypic Effects on Seeds During Development

Similar results have been observed in the desert species *Aegilops kotschyi* (Wurzburger and Koller 1976).

Photothermal effects have also been found in non-desert plants such as *Chenopodium album* L. (Chenopodiaceaea), a pluriregional species and a common weed of irrigated crops (Zohary 1966; Holm et al. 1977). Germination was higher both in light and in dark in seeds matured at 22/12 °C and under 8 h-SDs from flower bud formation, in comparison with seeds matured at 22/12 °C and in 18 h-LDs (Karssen 1970).

2.4.2 Daylength Affecting Seed Germination of Plant Species with Dry Fruit

2.4.2.1 Short or Long Day Effects

1. *Ononis sicula* Guss. (Fabaceae), is an annual species of the S. Mediterranean, Saharo-Arabian and Irano-Turanian geographical regions (Zohary 1972). In this plant, flowering is controlled facultatively by photoperiod: 8 h to 12.5 h light have a SD effect (Table 19) (Gutterman and Evenari 1972). Day length and water stress affect the first flowering node on the main branch, but the first harvest of mature seeds is affected only by the day length (Table 20). Under greenhouse conditions, the difference between SD

Table 19. Effect of photoperiod on the mean age and leaf numbers of *Ononis sicula* at the time of the first appearance of flower buds. (After Gutterman and Evenari 1972)

Photoperiodic treatment	Age at appearance of flower buds (days)	Number of leaves at appearance of flower buds
LD (20 h)	53.3	13
SD (8 h)	102.5	150
ND (11 – 12.5 h)	100.0	145

Table 20. Environmental effects on the onset of flowering of *Ononis sicula* and on the timing of the seed harvest. Plants sown 29 January 1963. (After Gutterman 1966)

Treatment	First flowering node (± S.E.)	Harvest (days from planting)			
		First	Main	Last	First to last harvest (days)
Greenhouse SD (8 h)	19.0 ± 0.0	126	146	150	24
Greenhouse LD (20 h)	7.7 ± 0.5	78	86	131	53
Outdoors SD (8 h)	7.0 ± 0.7	115	149	149	34
Outdoors LD (20 h)	4.8 ± 0.3	89	103	115	26

Environmental Effects

Table 21. Effects of the environment of the mother plant on the colour and weight (mg ± S.E.) of seeds of *Ononis sicula* at time of the main harvest (see Figs. 24 – 28). (After Evenari et al. 1966)

Treatment	Seed colour	Seed weight (mg)
Greenhouse SD (8 h)	Yellowish	0.57 ± 0.03
	Greenish	0.51 ± 0.4
	Brown	0.35 ± 0.02
Greenhouse LD (20 h)	Yellow	0.61 ± 0.03
	Greenish	[a]
	Brown	[a]
Outdoors SD (8 h)	Yellow	[a]
	Greenish	[a]
	Brown	0.23 ± 0.04
Outdoors LD (20 h)	Yellow	0.62 ± 0.03
	Greenish	0.53 ± 0.04
	Brown	0.23 ± 0.05

[a] Not present in the main harvest.

and LD for the first harvest is 48 days, but under outdoors conditions only 26 days. The longest period of seed maturation (from the first to the last harvest) is longer in LD (53 days) and shorter in SD (24 days) under greenhouse conditions (Table 20) (Gutterman 1966). Various treatments of the mother plants during seed maturation affect the germinability of seeds, through changes in the development of the seed coat and its surface structure. These modify its permeability to water, resistance against fungus, and the longevity of the seed. Under day length of 14.5 to 20 h (which has an LD effect), yellow seeds developed with well-developed seed coat surface structures. Under day length of 8 to 11 h (which has an SD effect), either brown seeds developed with undeveloped seed coats, and/or green or yellowish seeds with intermediate seed coat surface structure and greater water permeability. Day length during the last stage of seed maturation is the critical factor (Gutterman and Evenari 1972). The greenish, yellowish and yellow seeds differ in their germinability according to the day length and their weight and period of storage (Table 21; Figs. 21 – 23) (Gutterman 1966). Scarification of the seed coats of the yellow, yellowish, green and brown seeds, which also differ in surface structure (Figs. 24 – 28) brings their germination to 100% within 24 h (Gutterman and Evenari 1972; Gutterman 1973; Gutterman and Heydecker 1973).

2. *Trigonella arabica* Del. (Fabaceae) is an annual desert species and a good pasture plant of the Saharo-Arabian geographical region (Zohary 1972). Yellow seeds or yellow seeds with green spots are developed during maturation under LDs, and the green seeds or brown seeds are developed during maturation under SDs. The seed coat structure is better developed in the less permeable yellow seeds than in brown seeds whose seed coats are more permeable (as in *O. sicula*). (Figs. 29 and 30; Table 22) (Gutterman 1978a).

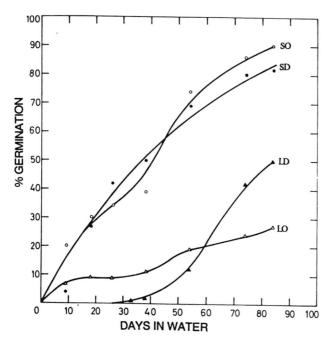

Fig. 21. Time course of germination of intact seeds of *Ononis sicula* grown in a greenhouse with photoperiods of 20 h (▲) and 8 h (●) and also of intact seeds grown outdoors with photoperiods of 20 h (△) and 8 h (○)

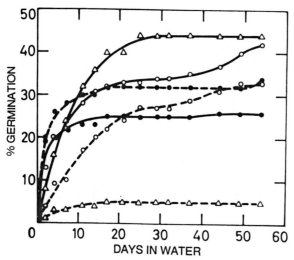

Fig. 22. Germination of the greenish (———) and yellow (– – –) fractions of a seed population harvested from mother plants grown in the greenhouse under a photoperiod of 8 h (●) and for similar fractions of the seed population grown outdoors under photoperiods of 8 h (○) and 20 h (△). Seeds grown in the greenhouse under a photoperiod of 20 h did not germinate. Seeds were tested 2 years after the first experiment. (After Evenari et al. 1966)

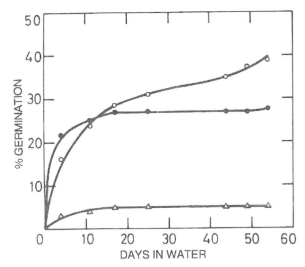

Fig. 23. As for Fig. 21, but for seeds tested 2 years later (greenish and yellow seeds only). (After Evenari et al. 1966)

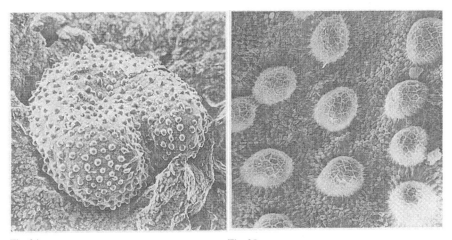

Fig. 24 **Fig. 25**

Fig. 24. Whole seeds of *Ononis sicula* (×49). (After Gutterman 1973; Gutterman and Heydecker 1973)

Fig. 25. Detail of yellow seed of *Ononis sicula* (×336) showing protuberances 'hills' on the surface. (After Gutterman 1973; Gutterman and Heydecker 1973)

These variations of the seed coat are dependent on the day length during the last 8 days of maturation when the fruit has reached its final size but is still green, as in *Ononis sicula* (see above). At this stage, whilst still on the mother plants, seeds matured in fruit covered with aluminium foil, and

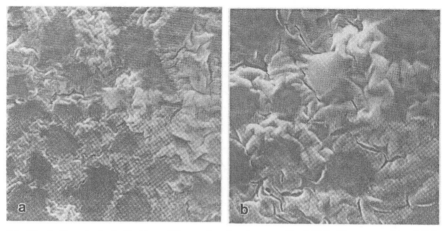

Fig. 26a, b. *Ononis sicula.* Section of the seed surface between the 'hills' of brown seed (a×3186; b×6372). Relatively thin and smooth surface covering an undeveloped structure (Gutterman 1973; Gutterman and Heydecker 1973)

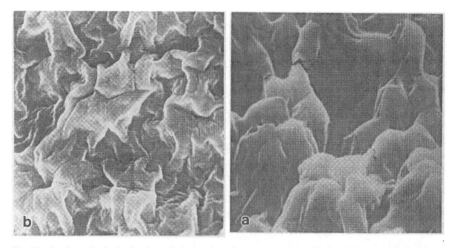

Fig. 27a, b. *Ononis sicula.* Section of the seed surface between the 'hills' of green seed (a×3248; b×6496). Surface structure well developed (Gutterman 1973; Gutterman and Heydecker 1973)

in uncovered fruits under 8 h-SDs and 15 h-LDs. Seeds from both covered and uncovered pods under 8 h-SDs swell completely after 3 weeks of imbibition. The seeds that matured on plants under 15 h-LDs, from uncovered and covered fruit, swell only 10 to 29% respectively (Table 23; Fig. 31). Scarification of the seed coats induces 100% germination within 24 h.

Seeds of *Ononis sicula* or *Trigonella arabica* that mature on a young plant under long days have a well-developed seed coat and are hard seeds

Environmental Effects

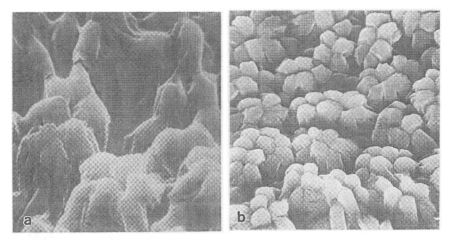

Fig. 28a, b. *Ononis sicula.* Section of the seed surface between the 'hills' of yellow seed (b×3045; a×6960). Well-developed and thickened surface structure. (Gutterman 1973; Gutterman and Heydecker 1973)

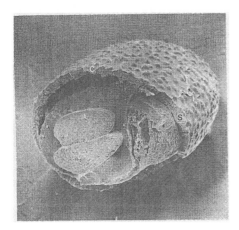

Fig. 29. *Trigonella arabica.* Seed cut (S.E.M.×40). (*S*) seed coat; (*C*) surface structure. (After Gutterman 1978a)

(impermeable to water). These seeds are possibly *long-term* seeds of the 'seed bank' in the soil. The seeds that mature on the same mother plants during short days have coats which are more water permeable, as well as seeds that have matured under long days during plant senescence. They are both the *short-term* seeds of the seed bank (Table 24) (Gutterman 1966, 1973, 1978a; Gutterman and Evenari 1972).

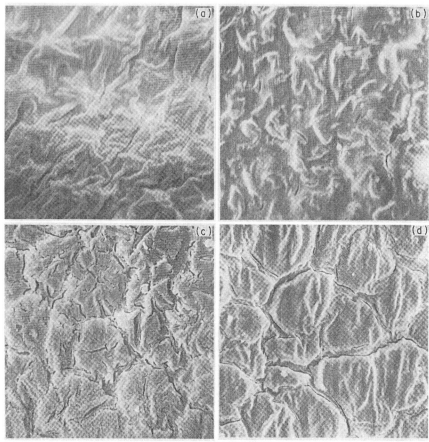

Fig. 30. *Trigonella arabica.* (S.E.M. ×3132); *a* Brown seed surface covering an undeveloped structure; *b* Surface structure of a green seed; *c* Surface structure of a yellow seed with green spots; *d* Well-developed surface structure of a yellow seed. Brown and green seeds developed on young plants under SD. Under LD yellow seeds or yellow with green spots develop. (After Gutterman 1978a)

Table 22. Colour and sweling capacity of seeds of *Trigonella arabica* ripened under photoperiodic regimes of 8 h SD or 20 h LD. Percentages of swelling after 3 weeks in water at 20°C (Fig. 30). (After Gutterman 1978a)

Photoperiod	Predominant seed colour	% Seeds that swelled[a]
SD	Brown	98
SD	Green	92
LD	Yellow with spots	30
LD	Yellow	4

[a] The non-swollen seeds failed to take up water even after 100 days.

Environmental Effects

Table 23. Influence of 8 h SD or 15 h LD on the permeability of the seed coat of *Trigonella arabica* during the final 8 days of seed maturation on the mother plant, when the fruit was covered by aluminium foil, or uncovered in the control (see Fig. 31). (After Gutterman 1978a)

Day length	Seeds that swelled after 3 weeks in water (%)	
	Fruit uncovered during ripening (control)	Fruit covered with aluminium foil during ripening
SD	100	100
LD	10	29

Fig. 31. *T. arabica* plant in bloom and with fruits (pods). On the day when plants were transferred from LDs (15 h) to SDs (8 h) or vice versa, the clusters of pods that had reached their final size but were still green were covered by aluminium foil. The others were signed and remained uncovered. (After Gutterman 1978a)

Table 24. Distribution of yellow, greenish and brown *Ononis sicula* seeds in the population, as affected by age at harvest and by the environment of the mother plant. Mother plants were planted on 29 January 1963 (Figs. 24 – 28). (After Gutterman 1966)

Environment	Plant age at harvest (days)	Distribution of seed population		
		% Yellow seeds	% Green seeds	% Brown seeds
LD greenhouse	78	100	1	0
	131	86	4	10
LD outdoors	89	97	2	1
	115	64	18	18

2.4.2.2 Quantitative LD Effect

As day length increases during seed maturation, the percentage germination of seeds is higher. This quantitative LD effect on seed germination is found in *Polypogon monspeliensis* and *Carrichtera annua*.

1. *Polypogon monspeliensis* L. Desf. (Poaceae) is an annual species from Mediterranean, Irano-Turanian and Saharo-Arabian regions, tropical and South African regions (Feinbrun-Dothan 1986). Plants from seeds harvested in Wadi Zin near Sede Boker were grown under six different day lengths outdoors as well as in the greenhouse. In seeds harvested from both, the longer the day length the higher the percentage of germination at 25 °C in continuous light (Table 25) (Gutterman 1982b).
2. *Carrichtera annua* (Cruciferae) is an annual desert plant from the Saharo-Arabian geographical region (Zohary 1966). After 9 years of dry storage,

Table 25. The influence of different day lengths, under greenhouse and outdoor conditions, on *Polypogon monspeliensis* during growth and seed maturation and on seed germinability. The photoperiodic treatments began on 20 – 24 December 1979 and were continued until the harvesting of the seeds in March 1980. Germinated (4×50 seeds) at 25 °C in the light in October 1980. (After Gutterman 1982b)

Day length during growth of mother plant and seed maturation (h)	Seeds from greenhouse plants Germination (%)		Seeds from outdoor plants Germination (%)	
	3 days	7 days	3 days	7 days
9.0	0.0	0.0	1.5	10.0
11.0	0.0	0.0	9.5	17.0
12.0	17.5	19.0	64.0	86.5
13.5	38.0	44.5	98.5	98.5
15.0	60.5	66.0	97.0	98.5
18.0	90.5	91.0	98.0	99.0
Control natural day length	93.0	96.0	91.5	93.0

Environmental Effects

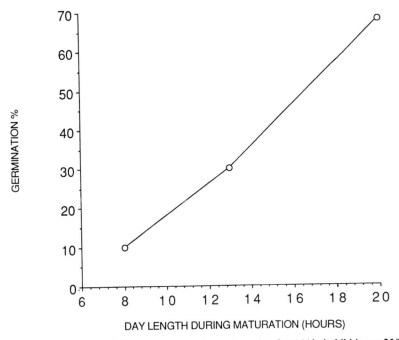

Fig. 32. Germination of *Carrichtera annua* L. Asch. seeds (after 144 h, imbibition at 25 °C in light) from covered fruits which matured during their final days in day lengths of 8, 13 and 20 h. Seeds germinated 2 years after harvest. (After Gutterman 1978c)

seeds matured on plants under 20 h-LDs show much better germinability (51.5%) than seeds from 8 h-SDs (7% germination), when germinated for 26.5 h at 25 °C in light (Gutterman 1973). Potted plants were transferred from natural day length of 13 h to either 8 h-SDs or 20 h-LDs, after the green fruit, which had reached full size, was covered with aluminium foil. Two years after full maturation, germination of the seeds that had matured in covered fruit at three day lengths, 8, 13 and 20 h, was compared at 25 °C in light after 144 h of imbibition. The longer the day length had been, the higher was the percentage of germination (Fig. 32) (Gutterman 1978c). As found also in *T. arabica* (Gutterman 1978a), the photoperiodic effect on seeds after the fruit had been covered is mediated via the leaves of the mother plants.

From these results, it is possible to speculate that *Carrichtera annua* seeds which matured under LDs in the late spring are those that will germinate first, while the seeds that matured under SDs in the early winter will remain in the 'seed bank' for an extended period.

2.4.2.3 Quantitative SD Effect

1. *Portulaca oleracea* L. (Portulacaceae) is an annual species of the warm-temperate regions of both hemispheres (Zohary 1966). It is one of the eight

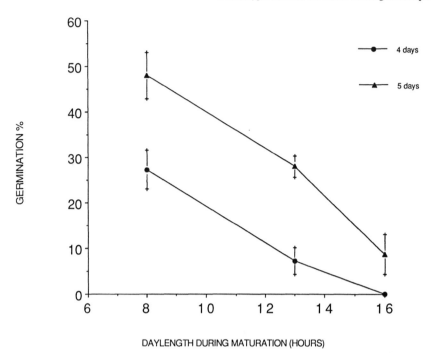

Fig. 33. The influence of a change in day length during the final 8 days of maturation of *Portulaca oleracea* L. seeds, from 16 h to either 13 h or 8 h days on germination at 40 °C in the dark with 5 min of white light once in 24 h. The first illumination was given 1.5 h after wetting (4 batches of 50 seeds each per treatment). Germination (% ± S.E.) after 4 or 5 days. (After Gutterman 1974)

most common weeds on earth (Holm et al. 1977). One of its centres of diversity is in Mexico, while numerous forms, which are different from those in all other parts of the world, are found in the deserts of Australia. The diploid sub-species of the New World grows mainly as an halophyte in coastal areas. The others, which are tetraploids, are weeds and have a wider range of latitude and altitude. The hexaploids are found mainly in high latitudes and altitudes (Danin et al. 1978). A higher percentage of seeds which matured on plants grown under 8 h-SDs germinate faster than seeds from 16 h-LDs (Gutterman 1974). This effect of day length is a quantitative, SD effect. The critical time is the last 8 days of seed maturation. When plants were transferred from 16 h-LDs to shorter days of 13 h or 8 h during the last 8 days of maturation, they were found to have different germinability. The shorter the day length, the higher the percentage of germination (Fig. 33) (Gutterman 1974).

Chenopodium polyspermum L. is an annual plant of waste-land and cultivated ground in the Euro-Siberian, Mediterranean and Irano-Turanian regions (Zohary 1966). The shorter the day length (from 24 h to 10 h), the higher the degree of germination, and the heavier the seeds. The thickness of the seed coat is 20 μ in seeds that matured under 8 h-SDs, compared with

Environmental Effects 53

46 µ for seeds matured under 24 h-LDs. Scarification of the seed coat brings germination to 100% (Jacques 1968). The critical time affecting seed germinability under 9 h-SDs is 8 days from the appearance of microscopically visible flowers, but for the LD effect 24 days are necessary. The seed coat thickness is quantitatively dependent on the number of the long days after the flowers begin to appear. The more LDs experienced, the thicker the seed coat (Pourrat and Jacques 1975).

2.4.2.4 The Influence of the Day Length on the Seed Coat

Seed coat thickness was found to be quantitatively affected by day length in *Chenopodium polyspermum*. The more LDs during the 24 days from flowering to seed maturation, the thicker the tegument (Jacques 1957, 1968; Pourrat and Jacques 1975). This has also been seen in *C. album* (Karssen 1970). In *C. bonus-henricus* the altitude was found to affect also the thickness of the seed coat (Dorne 1981) (see Sect. 4.2.1).

Seed coat structure. In *Ononis sicula* and *Trigonella arabica*, the seed coat is more developed under LD, and is less permeable to water than under SDs (Tables 22, 23; Figs. 21, 24–30) (Gutterman 1973, 1978a).

2.4.3 Day Length Affecting Seed Germination of Species with Soft Fruit

The tomato, Lycopersicum esculentum Mill. (Solanaceae), originates from S. America. Tomatoes are not desert plants, but the influence of day length is similar in them to that found in desert plants. This is a natural day length (ND) plant as regards flowering, but day length has been found to affect seed germination (as in *Carrichtera annua*). Under 6 h-SDs in over-ripe fruit, 38% of 297 seeds were found to germinate in the fruit in comparison with only 0.2% of 562 seeds that germinated when the plant had been under a day length of 13 h. None of 545 seeds were found to germinate in the fruit of plants under 20 h-LDs. When the fruit was covered and the plants transferred from one day length to another, almost the same results were seen as in *Trigonella arabica* (Gutterman 1978a) (Table 26). A much higher percentage of tomato seeds harvested from plants under 6 h-SDs germinated than of seeds from 20 h-LDs. The differences in germination were found to be even greater when the fruit was covered with aluminium foil. The day length during fruit maturation also has an influence on the inhibitory effect of the juice of the tomato fruit. When lettuce achenes imbibed half-strength tomato juice from fruit matured under 6 h-SDs, the percentage of germination was much higher (58%) than the percentage (8%) of germination in juice from fruit matured under 20 h-LD. In these, the inhibitory effect on seed germination was much higher (Table 26). When they are separated from the mother plant, these soft fruits respond to day length which affects the germinability of the seeds (Gutterman 1978b). In this situation the fruit responds as it would on the mother plant.

54 Phenotypic Effects on Seeds During Development

Table 26. (A) Effect of day length during period of fruit ripening on seed germination inside over-ripe tomato fruits. (B) Effect of day length on germination in petri dishes of tomato seeds from fruits uncovered (control) and covered with aluminium foil during maturation. (C) Percentages of lettuce achenes that germinated in half-strength tomato juice in petri dishes. (After Gutterman 1978b)

Photoperiodic treatments of the tomato plants (h)	A	B		C
	Tomato seeds % germination with fruits	Tomato seeds % germination in water from:		Lettuce seeds % germination in tomato juice
		covered fruit	non-covered fruit	
6	37.7	50	34	58
13	0.2	–	–	–
20	0.0	8	11	8

Table 27. (A) Effect of photoperiodic treatment on 4 (100 g) harvested pink tomato fruit of the release of ethylene (ppm) over 2 h, into a volume of 300 ml of gas above the fruit. (B) Effect of photoperiod and application of 0.5 ml of 2000 ppm ethephon on ethylene release from tomato fruits (ppm). (After Gutterman 1978a)

Part	Number of photo-periodic treatments	Ethylene release (ppm)		
		6 h SDs	9 h SDs	17 h LDs
A	1	5.9	6.8	8.9
	2	5.5	7.0	8.1
	3	3.6	6.5	7.0
	4	4.0	4.2	6.5
B	1	5.4	8.6	10.7
	2	12.0	11.3	17.1
	4	18.3	23.3	27.1

The amounts of ethylene released were found to be affected quantitatively by day length in harvested tomato fruits exposed to 6, 9 or 17 h of day length during ripening and seed maturation. The longer the day length, the larger the quantities of ethylene released from the fruit, even after a single exposure to the photoperiod. This phenomenon persisted for a number of days, until the fruits were fully ripe. The application of ethepon to tomato fruits during storage under the different photoperiods showed that ethepon enhances the role of day length in ethylene release (Table 27). These treatments also have an effect on the germinability of seeds harvested from fruits treated with ethepon under different day-length regimes (Gutterman 1978b).

Cucumus prophetarum is a perennial desert plant inhabiting the Saharo-Arabian and extending into the Sudanian geographical region (Feinbrun-Dothan 1978). The germination of seeds from ripe, turgid, juicy fruit harvested and

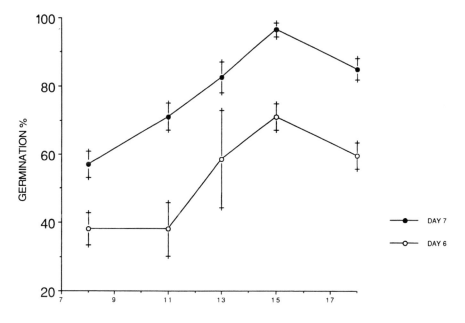

Fig. 34. Germination (% ±S.E.) of *Cucumis prophetarum* seeds after 6 days and 7 days of imbibition at 20 °C in continuous white light of seeds that were separated from juicy fruit stored for 9 days in 8, 11, 13, 15 and 18 h of light per day. Fruits were exposed to 8 h natural environmental conditions and, between 16 and 08 h, the chambers were covered and the fruit received incandescent and fluorescent light according to the total day length of each treatment (at the intensity of 15–20 µmol m^{-2} s^{-1}). The control under natural daylength (11.5 h) germinated to 66.5% ±3.8 after 6 days and 87.5% ±1.7 after 7 days (±Standard Error). (After Gutterman 1992d)

stored under different day lengths (8, 11, 13, 15 and 18 h) for 9 days, were found to have been influenced by the day length. After 6 days of imbibition, seeds from treatments of 8 and 11 h showed a lower level of germination than with longer day lengths. After 7 days of imbibition, a more gradual effect was seen, in which, with one exception, the longer the day length the higher the percentage of germination (Fig. 34) (Gutterman 1992a).

The cucumber, Cucumis sativus L., is a tropical annual plant originating from the Himalayas and probably taken into cultivation in northern India (Zohary and Hopf 1988). The storing of fruit after harvesting for 15 days under different photoperiodic regimes showed an influence on the germination of the seeds even after these had been kept for 270 days in dry storage. The seeds were allowed to imbibe water at 20 °C in the dark. After 24 h imbibition, 84% of seeds from 8 h SDs germinated, but only 12.5% from fruit held under LDs (Table 28) (Gutterman and Porath 1975; Gutterman 1978b). Injection of ethephon, which releases ethylene, into the soft fruit of cucumbers can also

Table 28. Germination (% ± S.E.) (N = 200:4×50) of *Cucumis sativus* seeds (after 24 h imbibition at 20°C in the dark) as influenced by day length and ethephon treatment during 15 days' post-harvest storage of the ripe fruit. The dry seeds had been stored for 270 days in the dark before the experiment. (After Gutterman 1978a)

Chemical to the fruit	Day length and germination (%)	
	SD (8 h)	LD (20 h)
Control	84 ± 5.9	12.5 ± 8.2
Ethephon 1000 ppm	2 ± 2	79.0 ± 3.4

affect germinability in a way opposite to day-length treatments. From seeds matured under SDs, 2% germinated while 79% germinated from seeds matured under LDs (Table 28) (Gutterman 1987b).

When GA 4–7, 100 ppm, was injected into the cucumber fruits which had reached their final size and been harvested, there was an increase in germinability. The opposite effect was found when ABA was injected into the cucumber fruit: none of the seeds germinated (Gutterman and Porath 1975; Gutterman 1978b).

From studies on other plants species, it is well known that hormone application to plants during maturation has an influence on seed germinability (Black and Naylor 1959; Zeevart 1966; Jackson 1968; Felippe and Dale 1968; Gutterman et al. 1975; Ingram and Browning 1979).

2.4.4 The Critical Time for the Day-Length Effects

In *O. sicula*, *T. arabica*, (Fabaceae) (Table 23) (Gutterman and Evenari 1972; Gutterman 1978a), *Lactuca sativa* (Compositae) (Table 29) (Gutterman 1973),

Table 29. Germination of achenes of *Lactuca sativa* Grand Rapids 517 (% ± S.E.) (N = 200:4×50) at 26°C in darkness with short periods (5 min) of white light; after photoperiodic treatment (8 h SD and 16 h LD). (After Gutterman 1973)

Photoperiodic conditions during growth of mother plants	Germination (%) after 2 days	Germination (%) after 11 days
SD, then LD (80)[a]	0	4 ± 1.4
LD, then SD (80)	29.5 ± 2.6	32 ± 3.2
SD, then LD (12)	5.0 ± 1.2	8 ± 1.4
LD, then SD (12)	13.5 ± 3.4	18 ± 3.4
continuous LD	5.0 ± 0.6	6 ± 0.8
continuous SD	16.5 ± 2.2	24 ± 2.5

[a] Number of days under the photoperiodic conditions before harvest. The germination experiment began immediately after harvest.

Environmental Effects 57

Portulaca oleracea (Portulacaceae) (Fig. 33) (Gutterman 1974) and *Carrichtera annua* (Cruciferae) (Fig. 32) (Gutterman 1978b, c), the critical time for day-length effects is during the last stages of seed maturation – from when the fruit has reached its final size but is still green, to full maturation 7 to 14 days later (Gutterman 1978a). In soft fruit, such as cucumbers, 5 to 15 days are also sufficient for the day-length effect (Table 28) (Gutterman and Porath 1975; Gutterman 1978b). This was found to be the same in tomatoes (Gutterman 1978b). After the transfer of plants from LDs to SDs or from SDs to LDs 12 days before the harvest of the achenes of lettuce (*Lactuca sativa*), the last day-length treatment affected germination similarly to the SDs or LDs controls, respectively (Table 29). In *Chenopodium album* L., transfer from LDs to SDs at the time of flower bud formation increased the percentages of germination both in the light and in the dark at 22/12°C and at 22/22°C, in comparison with continuous LDs (Karssen 1970). In *C. polyspermum*, 8 days after flower bud formation is the critical time affecting seed germination in 8 h SDs, but in 16 h LDs 24 days are needed (Pourrat and Jacques 1975).

2.4.5 Natural Conditions of Maturation During Summer and Autumn Affecting Seed Germination

Lactuca serriola L. (Asteraceae), is a widely spread annual (or biennial) species of the Mediterranean, Irano-Turanian, Europe-Siberian and Sudanian areas extending into South Africa (Feinbrunn Dothan 1978). It is an LD plant in respect to flowering (Gutterman et al. 1975).

Ripe achenes ('seeds') were collected each month from July to October 1989, from a natural plant population near Sede Boker. They differed in weight and in germinability when tested in October and December 1989, and January, March and May 1990, under constant temperatures with cycles of light and dark (Figs. 35–37; Table 30). The germination percentage of *L. serriola* achenes harvested in August and germinated in September 1989 showed light sensitivity at 10 and 30°C but not at 15 and 20°C (Fig. 38). The achenes of the different harvests showed different relative germinability under temperatures between 10 and 25°C in the light and in the dark. This was seen, after 24 h and 48 h of imbibition, in achenes harvested during July and October (Fig. 39). Imbibition during the following winter and spring, and 1 year later, under natural temperatures in the range of 2.5–28°C, and natural day length or in dark, showed significant differences in the germination rates of the different harvests (Figs. 40, 41) (Gutterman in press). In this area, January is usually the month with most rainfall and is the main time when germination occurs. The plants began to produce seeds from the first capitula in July, when the photoperiodic day length is the longest (15 h), and ceased to do so in October, when the day length is much shorter (12 h) and when the majority of the leaves are in senescence.

The maturation of seeds with different germinability, which could be expressed under different temperatures and dates after maturation, has a very

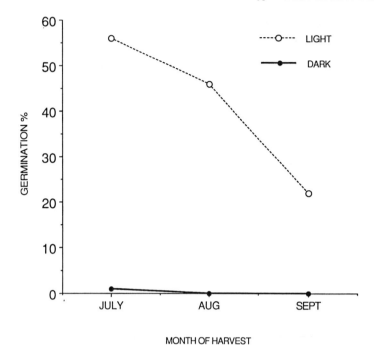

Fig. 35. Germination (%) of *L. serriola* achenes after imbibition at 26°C for 24 h in dark and light. Achenes harvested during July, August and early September 1989 and tested on 13 October 1989. (After Gutterman 1992c)

Table 30. Month of harvest of *Lactuca serriola* achenes: mean achene weight (±S.E.) (N = 200 : 4 × 50) and range of day length during maturation. (Gutterman 1992c)

Month of harvest 1989	Achene weight		Day length
	Mean weight (mg) of 200 (×5) achenes	Weight of 1000 achenes (mg)	Photoperiodic day (h)[b]
June	–	–	15 h$^{00'}$
July	89.36 ± 0.64	446.8	14 h$^{25'}$
August	89.56 ± 0.44	447.8	13 h$^{35'}$
Sept. 1–15	89.52 ± 1.0	447.6	13 h$^{10'}$
Sept. 16–30	99.98 ± 0.87[a]	502.9	12 h$^{45'}$
October	106.50 ± 0.26[a]	532.5	12 h$^{00'}$

[a] Significant difference by Fisher and Scheffe F-test at 95%. ANOVA test p = 0.0001.
[b] The day length mentioned is on the 21st of each month, with the exception of 1–15 September.

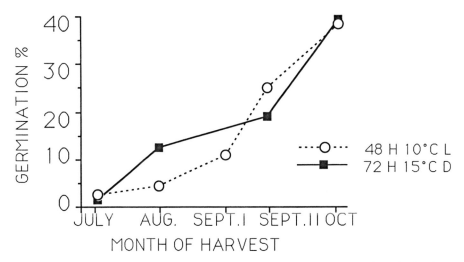

Fig. 36. Germination (%) of *L. serriola* achenes harvested from July to October 1989 after imbibition at 10 °C for 48 h in light and tested on 25 March 1990; or after imbibition at 15 °C for 72 h in the dark and tested on 6 May 1990. (After Gutterman 1992c)

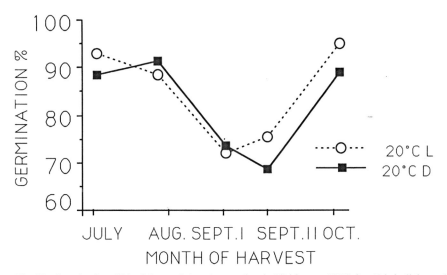

Fig. 37. Germination (%) of *L. serriola* achenes after imbibition at 20 °C for 48 h in light and dark. Achenes harvested from July to October 1989 and tested on 21 January 1990. (After Gutterman 1992c)

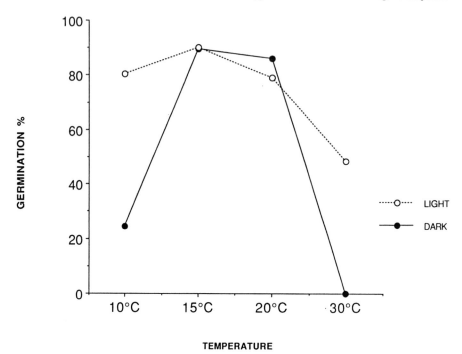

Fig. 38. Germination (%) of *L. serriola* achenes after 48 h at temperatures from 10–30 °C, in the light and in the dark, harvested during August and tested on 10 September 1989

important ecological advantage for the survival of this plant species under desert conditions by spreading the germination of the seed bank population over time.

2.4.6 Natural Winter, Spring or Summer Conditions of Maturation Affecting Seed Germinability

2.4.6.1 Spring Maturation

Natural day length was found to affect hard-seededness. In subterranean clover (*Trifolium subterraneum*), the longer the growing season in the spring in West Australia, the larger the proportion of hard seeds (Aitken 1939; Quinlivan 1965). Exposure of the seeds to wide soil surface temperature fluctuations during the dry summer increases the degree of softening of hard seeds (Sect. 4.2.1.3).

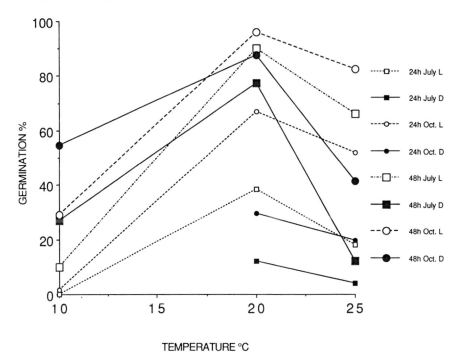

Fig. 39. Germination of *L. serriola* achenes after 24 h and 48 h of imbibition at 10, 20 and 25 °C, in light and dark. Achenes from two harvests: July and October 1989 and tested on 21 May 1990. (After Gutterman 1992c)

2.4.6.2 Winter or Summer Maturation

In bi-seasonal flowering perennial shrubs of Aizoaceae from South Africa that were introduced and grown in the Negev Desert highlands, large differences in seed germinability according to the season of seed maturation were found in natural day length and temperatures. Seed maturation in winter, when days are short (the shortest photoperiodic day length in the Negev desert highlands is about 11 h on 21 December) and temperatures are low, or in summer, when days are long (the longest photoperiodic day length is about 15 h on 21 June) and temperatures are high, seems to influence these differences in these perennial plants (Fig. 42; Table 31). When germination was tested at various temperatures, levels of germination differed at different temperatures after 100 days of imbibition (Figs. 43–45) (Gutterman 1991).

2.4.7 Altitudinal Effects

Seeds of *Chenopodium bonus-henricus*, collected from plants in natural populations at an altitude of 600 m, showed a higher percentage of germination

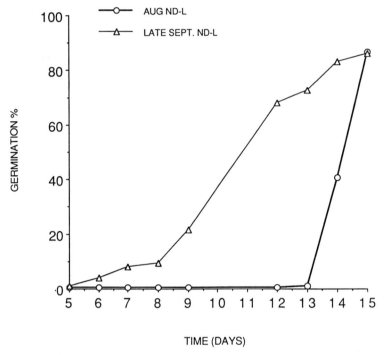

Fig. 40. Germination (%) of *L. serriola* achenes as a function of time (days). Achenes from August and late September 1989 harvests, after imbibition under natural conditions of temperatures 3–28 °C) and day lengths (in shadow) at Sede Boker (see text). Tested on 13 February 1990. (After Gutterman in press)

than did seeds collected from plants at an altitude of 2600 m. The higher the altitude, the lower the percentage germination and the thicker the seed coat, as well as the higher the polyphenol content of the seed coat. Plants transferred from one altitude to another produced seeds which were correlated with each altitude. The polyphenols that accumulated in the thicker seed coats of seeds which matured in plants at higher altitudes inhibited the germination. It is possible that increase in the polyphenol content is due to increased visible radiation at the higher elevation (Dorne 1981).

2.4.8 Influences of Light Quality During Maturation on Seed Germination

2.4.8.1 Under Artificial Laboratory Conditions

Arabidopsis thaliana seeds matured under white light with a high R/FR ratio show a higher percentage of germination in darkness than seeds from plants grown in light with a low R/FR ratio (McCullough and Shropshire 1970).

Environmental Effects

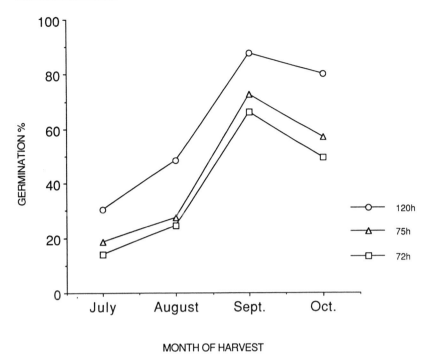

Fig. 41. *L. serriola* achene germination % after 72 h, 75 h and 120 h of imbibition under natural daylength and natural temperatures ranging from 2.5 °C to 20 °C. They matured and were harvested from July to October 1990. Tested on 7 January 1991. (After Gutterman in press)

The flowering and seed germination mechanisms of Portulaca oleracea. This plant shows a quantitative effect of day length during maturation on seed germinability (Fig. 33). Treatments of 2 h Red light, and then 14 h dark (D) followed by 8 h daylight during maturation engendered the greatest difference in seed germination when compared with 2 h Far-Red at the same time (56.5% and 14%, respectively). It had no effect, however, on flowering, while 2 h Red or Far-Red light in the middle of the dark period had the greatest influence on flowering (13.8 or 6 leaves at flower bud appearance, respectively) but none on seed germination (29.5% and 25%, respectively). The 2 h of Red or Far-Red treatment at the end of a dark period of 14 h had no effect on either flowering (5.9 or 6.7 leaves, respectively) or on seed germination (13% and 10.5%, respectively) (Table 32) (Gutterman 1974). It appears, therefore, that different light treatments during development and seed maturation may have different influences on the number of leaves before the appearance of the first flower bud, and on the germination percentages of seeds harvested. These varying responses in flowering and seed germination are even more pronounced after treatment with 8 h R or FR light before or after the 8 h D period, compared with 8 h in white light following 8 h of daylight. The 8 h R or FR following 8 h daylight has an SD effect, while 8 h R or FR following the 8 h D has a pro-

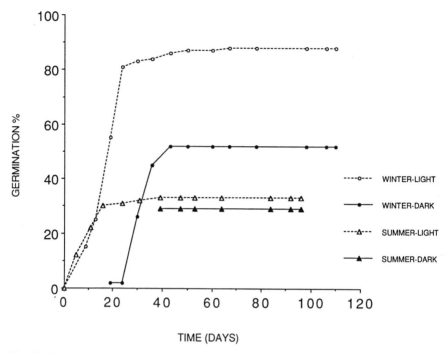

Fig. 42. *Cheiridopsis aurea* germination in light or dark at 35 °C of winter or summer matured seeds. (After Gutterman 1991)

Table 31. Germination (% ± S.E.) in the Light (*L*) and Dark (*D*) of seeds of *Cheiridopsis aurea* and *Cheiridopsis* No. 86 matured in winter (*W*) and summer (*S*). (After Gutterman 1991)

Season of maturation	Germination conditions	*Cheiridopsis aurea* 35 °C 50 days[a]	20 °C 13 days[a]	10 °C 50 days[a]	*Cheiridopsis* No. 86 35 °C 60 days[a]	20 °C 60 days[a]
W	L	92% ± 14.0	98% ± 1.0	55% ± 1.5	85% ± 1.5	18% ± 1
S	L	33% ± 6.5	79% ± 5.0	85% ± 1.5	37% ± 5.5	80% ± 0
W	D	52% ± 0	1% ± 0.5	98% ± 1.0	3% ± 0.5	62% ± 3
S	D	29% ± 0.5	98% ± 1.0	92% ± 4.0	7% ± 0.5	88% ± 1

[a] Days of imbibition at the final level of germination.

nounced LD effect on flower bud appearance. All of these four treatments have a SD effect as far as germination percentages are concerned. This is different from treatment with 8 h white light, which has a LD effect both on flowering and on germination (Table 32) (Gutterman 1974).

The photoperiodic response for germination was also found in two other species: *Carrichtera annua* and the tomato, *Lycopersicum esculentum*, both of which are day-neutral with regard to flowering. In these two species there is no

Environmental Effects

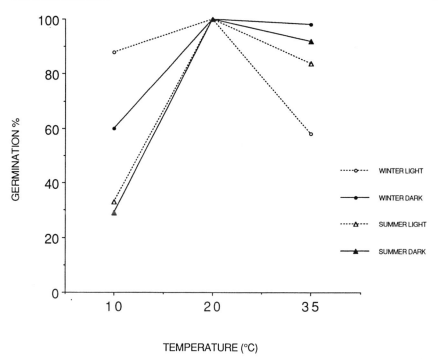

Fig. 43. *Cheiridopsis aurea* levels of germination, after 100 days of imbibition at 10, 20, and 35 °C, of winter or summer matured seeds. (After Gutterman 1991)

photoperiodic effect on flowering but there is an effect on seed germinability. From the above it would seem that the mechanisms affecting flowering are different from those that affect seed germinability — at least in the species mentioned, and under the conditions in which they were tested.

In the case of Cucumis prophetarum, storage of fruit for 30 days and, in the case of *Cucumis sativus*, storage for 5 days in continuous Red light, increases the percentage of Pfr of the photoreversible phytochrome in the seeds. Germination in darkness was 92–100% respectively, at 23–25 °C, after 50 h imbibition for *C. prophetarum* and 170 h for *C. sativus*. In fruits stored in continuous Far-Red light, all the photoreversible phytochrome of the seeds of both species was in the P_r stage. Germination of *C. sativus* was 27.5% in the dark. After the first 50 h imbibition at 23–25 °C no seeds of *C. prophetarum* germinated in the dark. Storage of *C. sativus* fruits in the dark give similar results to storage in FR, while sunlight gives similar results to R light (Table 33) (Gutterman and Porath 1975).

In another set of experiments ripe, juicy and turgid fruits of *C. prophetarum* were stored for 9 days under Red, Far-Red, light or dark conditions. Seeds separated from fruit which had received the various treatments imbibed water in white light, immediately after separation from the wet fruit. After im-

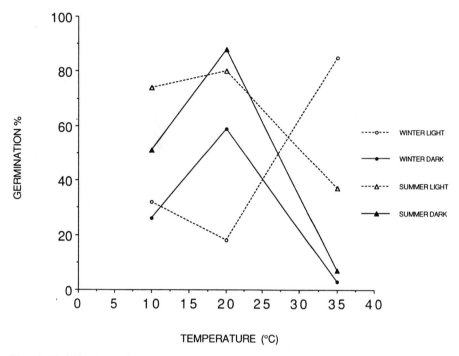

Fig. 44. *Cheiridopsis* (species No. 86) levels of germination after 100 days of imbibition at 10, 20, and 35 °C of winter or summer matured seeds. (After Gutterman 1991)

bibition in light, germination of seeds from fruit stored under dark and FR, was much higher than germination from fruits stored under R light (Fig. 46) (Gutterman 1992b).

The total photoreversible phytochrome is much higher after storage of the fruit in Far-Red light, compared with seeds that were taken from fruit stored in Red light. Seeds that contain more Pfr germinate to a higher level in the dark. When they germinate in light, however, the higher the level of photoreversible phytochrome, the higher the percentage of germination (Fig. 46) (Gutterman 1992b), (Table 33) (Gutterman and Porath 1975). This correlates well with the percentage of the Far-red absorbing phytochrome (% Pfr) in the total photoreversible phytochrome in the seeds (Table 33; Figs. 47–51). Seeds of *C. prophetarum* contain 30% H_2O, and of *C. sativus* 45% H_2O, when separated from the fruit. However, after the seeds were exposed to a period of dry storage, these differences in germination in the dark disappeared completely (Gutterman and Porath 1975).

In Chenopodium album, the Red light effect during seed maturation disappears 4 months after maturation (Karssen 1970).

Environmental Effects

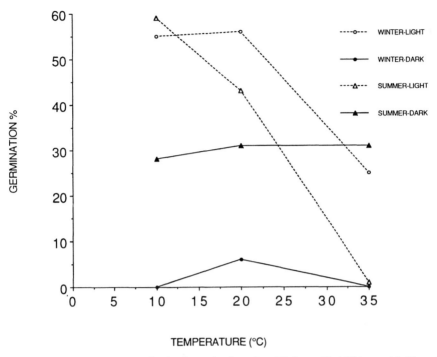

Fig. 45. *Juttadinteria proximus* levels of germination after 100 days of imbibition at 10, 20, and 35 °C of winter or summer matured seeds. (From Gutterman 1991)

Table 32. Photoperiodic conditions during growth of mother plants of *Portulaca oleracea* and seed maturation, showing effect on the number of leaves at the time of appearance of the first flower bud, and the seed germination (% ± S.E.) after 24 h of imbibition at 40 °C and white light. (*R*) Red light; (*FR*) Far Red light; *L* Daylight; *W* White light; *D* Dark (± S.E.). (After Gutterman 1974)

Photoperiodic conditions	No. of leaves at flower bud appearance	Germination (%)
2R, 14D, 8L	5.7 ± 0.2	56.5 ± 2.8
2FR, 14D, 8L	6.1 ± 0.3	14.0 ± 1.2
7D, 2R, 7D, 8L	13.8 ± 0.3	29.5 ± 7.0
7D, 2FR, 7D, 8L	6.0 ± 0.2	25.0 ± 1.4
14D, 2R, 8L	5.9 ± 0.3	13.0 ± 1.4
14D, 2FR, 8L	6.7 ± 0.2	10.5 ± 0.7
8D, 8R, 8L	15.8 ± 0.6	29.0 ± 3.0
8D, 8FR, 8L	12.1 ± 0.7	34.0 ± 2.6
8R, 8D, 8L	6.8 ± 1.0	25.5 ± 2.8
8FR, 8D, 8L	8.7 ± 0.5	28.0 ± 3.9
8W, 8D, 8L	11.3 ± 0.2	18.5 ± 1.7
16D, 8L	7.9 ± 0.2	30.0 ± 1.6

Table 33. Amounts of photoreversible phytochrome, state of phytochrome in the seeds, and germination (% ± S.E.) (at 25 °C) of *Cucumis prophetarum* (after 50 h) and of *C. sativus* seeds (after 170 h). Fruit stored under different light conditions (at 23 – 25 °C). (After Gutterman and Porath 1975)

Species	Light conditions on the harvested fruits	Photoreversible phytochrome in the seeds $(\Delta \cdot O \cdot D) \times 10^{-4}$	Far-red absorbing phytochrome in seeds (% Pfr)	Seed germination (%) in the dark
C. prophetarum	Red	8.6 ± 0.4	93.5	100.0
	Far-red	20.0 ± 0.7	0.0	0.0
C. sativus	Red	24.0 ± 2.1	53.3	91.9 ± 4.3
	Far-red	35.1 ± 2.0	0.0	27.5 ± 4.3
	Dark	36.4 ± 1.8	8.5	22.5 ± 2.5
	Sunlight	28.8 ± 2.0	47.2	100.0

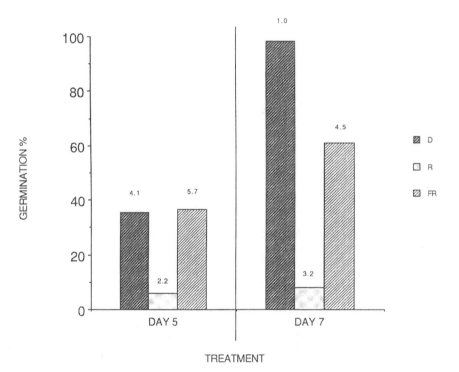

Fig. 46. Germination (%) of *Cucumis prophetarum* at 20 °C in continuous white light after 5 and 7 days of imbibition of seeds separated from juicy, turgid fruit and stored for 9 days under Far Red (*FR*), Red (*R*) and Dark (*D*) in laboratory conditions. The *numbers above the columns* represent the ±Standard Error. (After Gutterman 1992d)

Environmental Effects

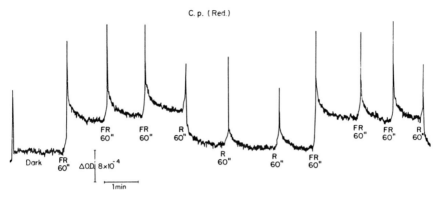

Fig. 47. Phytochrome transformation in a 10 mm cuvette filled with whole *Cucumis prophetarum* seeds (30% H_2O) removed from fruits stored in continuous Red light (2000 erg/cm² s) for 30 days. Relative optical density responses as checked by a dual wavelength spectrophotometer (730/806 nm, temperature). Seeds were irradiated for 1 min before recording began. Red ($R = 660$ nm) and Far Red ($FR = 728$ nm) actinic light ($10^4 - 10^5$ erg·cm² s). Transmission 2%. *Dark* base line without subsequent illumination (see text). (After Gutterman and Porath 1975)

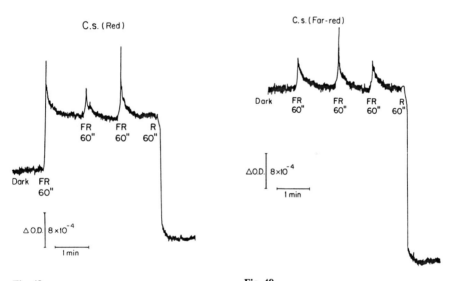

Fig. 48   Fig. 49

Fig. 48. Phytochrome transformation in *C. sativus* seeds (45% H_2O), removed from fruits stored in continuous Red light (2000 erg/cm² s) for 5 days. All other details as in Fig. 47. (After Gutterman and Porath 1975)

Fig. 49. Phytochrome transformation in seeds of *C. sativus* removed from fruits stored in continuous Far-Red light (2500 erg/cm² s) for 5 days. All other details as in Fig. 47. (After Gutterman and Porath 1975)

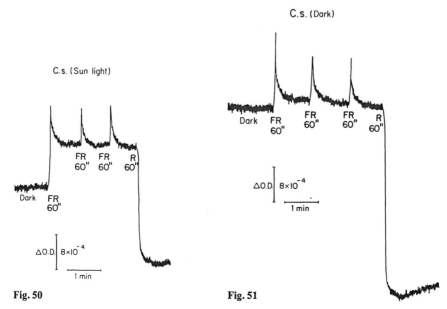

Fig. 50. Phytochrome transformation in seeds of *C. sativus* removed from fruits stored outdoors and receiving sunlight for 5 days. All other details as in Fig. 47. (After Gutterman and Porath 1975)

Fig. 51. Phytochrome transformation in seeds of *C. sativus* removed from fruits stored in continuous darkness for 5 days. All other details as in Fig. 47. (After Gutterman and Porath 1975)

2.4.8.2 Under Natural Conditions

The green leaf canopy was found to inhibit the germination of matured light-sensitive seeds (Black 1969; Vander Veen 1970; Gorski 1975; King 1975; Fenner 1980a, b). Only 1 hour under leaf-transmitted light is sufficient to inhibit germination of *Bidens pilosa* seeds in the dark (Fenner 1980b). During seed maturation on the mother plant there is in various species a relationship between concentrations of chlorophyll, which surround the developing seeds during maturation, and dehydration. In seeds that mature surrounded entirely by maternal green tissues, most of the phytochrome is arrested in the inactive form (Pr). These seeds, therefore, require a light stimulus for germination in the dark (Cresswell and Grime 1981). It is possible that this phenomenon may also occur in desert plants whose seeds need light for germination. But tests to determine whether this is the case have not yet been carried out.

2.4.9 Temperatures During Maturation Affecting Seed Germination

In *Aegilops geniculata* (*A. ovata*), all orders of caryopses are lighter and germinate to higher percentages after 24 h of imbibition at temperatures of

Environmental Effects

28/22 °C than caryopses that matured at temperatures of 15/10 °C (Datta et al. 1972a) (Table 16) (Sect. 2.2.3).

2.4.10 Conditions of Maturation and Different Levels of Germination Under Different Temperatures During Imbibition

After imbibition and transfer from low to high temperatures different levels of germination were observed in seeds of *Portulaca oleracea* that had matured under 2 h Red or Far-Red before the dark period (Table 34) (Gutterman 1974). In some species of Aizoaceae, seeds that have matured in different seasons (winter or summer) under natural conditions germinate to different levels after imbibition at different temperatures (Figs. 43–45) (Gutterman 1991). Achenes of *Lactuca serriola* which matured on different dates during summer showed the same phenomenon (Fig. 39). (Gutterman 1992c).

2.4.11 Water Stress During Maturation Affecting Seed Germination

Hirschfeldia incana (Cruciferae) is a plant from the Mediterranean and Irano-Turanian geographical regions. When the green wet seeds were soaked in water at 26 °C for 4–6 weeks after anthesis (WAA), they did not germinate. However, seeds removed 4–6 WAA and kept for 2 weeks at room temperature dried, remained green, and 91% germinated after inhibition in light at 26 °C. When plants with pods 4–6 WAA were not watered for 2 weeks, the seeds dried, their colour turned to brown and they germinated to 70% in light. Seeds 3–5 months after anthesis are brown and dry and germinated to 10% in light.

After being removed and desiccated, premature seeds germinated to much higher percentages than non-desiccated seeds or mature seeds. The desiccation or water stress of premature seeds prevents or decreases their after-ripening (Table 35) (Evenari 1965a).

Similar research has been carried out on non-desert plants. Immature developing seeds of soybean will not germinate after imbibing water unless

Table 34. Influence of Red (*R*) and Far Red (*FR*) light treatments to the mother plants of *Portulaca oleracea* L. During growth and seed maturation, on seed germination (% ± S.E.) after seed imbibition at temperatures of 25, 30 and 40 °C in continuous white light. (After Gutterman 1974)

Photoperiodic treatments to mother plants	Germination (%)		
	25 °C (6 days)	30 °C (6 days)	40 °C (1 day)
2R, 14D, 8L	2.0 ± 0.0	63.5 ± 5.5	98.5 ± 0.6
2FR, 14D, 8L	26.5 ± 1.8	63.0 ± 3.7	79.0 ± 2.4

Table 35. Germination of seeds of *Hirschfeldia incana* at different stages of maturation. Germination (%) in light (*L*) and in the dark (*D*). (Evenari 1965a)

Stage and germination (%)	Stage of maturation					
	I 4–6 weeks after anthesis	II As I, but fruits taken off, for 2 weeks in room	IIa As I, but plants not watered for 2 weeks	III 2 months after anthesis	IV 2–3 months after anthesis	V 3–4 months after anthesis
Colour of fruit coat	Green, wet	Green, dry	Brown, dry	Green-yellow, wet	Yellow-brown drying	Brown dry
Colour of seed coat	Green, wet	Green, dry	Brown, dry	Green, wet	Brown, wet	Brown, dry
Colour of embryo	Green, wet	Green, dry	Brown, dry	Green, wet	Cotyledons yellow-green, rootlet white, wet	Yellow-white, dry
Germination % at 26°C	0%	91% (L)	70% (L)	40% (L) 0% (D)	34% (L) 0% (D)	10% (L) 0% (D) Embryo germination 100%

Environmental Effects 73

they have previously been desiccated (Adams et al. 1983). However, soybean plants exposed to drought during seed fill show a lower standard germination percentage after a greater degree of stress (Dornbos et al. 1989).

Immature developing seeds of *Ricinus communis* (castor bean) and *Phaseolus vulgaris* do not germinate when removed from the capsules and soaked in water. When such seeds were removed and stored at a relatively high humidity their water content slightly declined and they germinated when soaked in water (Kermode et al. 1986; Bewley et al. 1989). Water stress during maturation has the ability to switch seed from the developing to the germinating system (Kermode et al. 1986). This includes changes such as protein patterns (Lalonde and Bewley 1986), messenger-RNA (Bewley et al. 1989) in the endosperm of *Ricinus communis* (Kermode et al. 1989a) and in other systems of the developing seed that are affected by desiccation, such as ABA (Kermode et al. 1989b).

2.4.12 Dimorphism and Achene Germination Dependent on Plant Size

In populations of *Heterotheca latifolia* (Compositae) found in Austin, Texas (30°18'N 97°47'W), large plants produce larger heads with relatively more disc achenes (which have a wind-dispersed pappus) than small plants of the same populations. These produce smaller heads with relatively more ray achenes which differ in their structure from the disc achenes, and do not have pappi. The embryo in the disc achenes are 60% heavier than the ray embryos. The ray achenes are dispersed over shorter distances from the plants and have a 'caution' or 'low-risk' strategy. In this germination is delayed and spread over a longer period. The disc achenes, which are wind dispersed by pappi, may germinate during the first available rain (Venable and Levin 1985).

2.4.13 Heterocarpy and Germinability

Two desert summer annual Chenopodiaceae, *Halothamnus hierochunticus*, inhabiting the West Irano-Turanian plant geographical region, and *Salsola volkensii* of the East Sahara Arabian plant geographical region (Zohary 1966), were found to produce heterocarpic seeds. *H. hierochunticus* develops a green and yellow 5-lobed fruiting perianth enclosing one embryo (a single-seeded fruit), which is a winged dispersal unit. The yellow perianth of *Salsola volkensii* is smaller (4–6 mm – including the wings). The green perianth is larger (5–8 mm) and the embryo consists of two green cotyledons. Freshly harvested green fruits germinate within 2 days, in light and dark to 96%, but the yellow fruits germinate to only 30% in light and 10% in dark after 16 days. The green fruit does not germinate after storage of more than 24 months. In contrast, the germination of yellow fruits increased with length of storage time, and after 5 years germinated to 73% in light and 16% in dark. Similar results were observed in the germination of the two-coloured fruit of *H. hierochunticus*

74 Phenotypic Effects on Seeds During Development

hierochunticus (see Chap. 5) (Koller and Negbi 1966). See also Kigel 1992, about heteromorphism of *Hedypnois cretica* (L.) F. W. Schmidt (Asteraceae).

2.5 Polymorphic Seeds and Germination

The annual halophyte *Atriplex triangularis* is widely distributed in inland salt marshes as well as in coastal areas of North America. Polymorphic germination of its seeds has been studied among populations in Ohio. Seeds were separated according to their diameter into three groups: (1) largest seeds, >2.0 mm; (2) intermediates, 2.0–1.5 mm; (3) smallest seeds, <1.5 mm. Salinity and temperature were found to affect the percentages of germination of each size group: 60% of the largest seeds germinated at 5°C night, 25°C day temperatures and 1.5% NaCl, but only 5% of the smallest did so (Khan and Ungar 1984a). The larger the seed, the higher is its salt tolerance (Khan and Ungar 1984b). The thickness of the seed coats of the three groups of seeds also varies. The single large sclereids of the seed coats have the larger diameter in the small seeds (21.5±0.37 mm); medium seeds have diameters of 16.8±0.45 mm, while the largest seeds have the thinnest layer (13.6±0.41 mm) (Khan and Ungar 1985).

Heteromorphism and different germination was found in Israel in *Hedypnois cretica* (Feinbrun-Dothan and Danin 1991) in populations from the Mediterranean (Jerusalem) to the desert (Arad) region (Kigel 1992) and in South Africa (Beneke 1991) (Sect. 2.2.7).

2.6 Conclusions

In many species, as far as seed germinability is concerned, the fate of the next generation is dependent on conditions of maturation when the seeds are still on the mother plant (Gutterman 1969). Maternal position and environmental factors affect differences in germinability during development and seed maturation. The last 5–15 days of seed maturation are critical. This has been confirmed in several species. Seeds showing different germinability develop on the same mother plant, as do those on plants of the same species but growing in different environments. If genotypic influences ensure fitness, seeds should germinate at the most propitious time in the right season and in the right place, but phenotypic influences during maturation ensure that, even under optimal conditions, only a portion of the population will germinate after one rain event or in one season.

In the desert species *Pteranthus dichotomus*, only one, or a maximum of two seedlings appear in one season from a dispersal unit containing seven "seeds" (Fig. 17) (Evenari 1963; Evenari et al. 1982). Similarly, in the Mediterranean *Aegilops geniculata* (*A. ovata*) in its natural habitat only one or two seedlings appear in one season, from five caryopses of a three-spikelet dispersal unit or six caryopses of a four-spikelet dispersal unit (see Fig. 18), (Datta

Conclusions

et al. 1970, 1972a, b; Gutterman 1992a). In *Ononis sicula* and *Trigonella arabica* (Tables 22, 23; Figs. 21 – 23) all the brown seeds with undeveloped seed coats swell during the rains of the following season. A small portion of the green and yellowish seeds also swell and germinate. The yellow seeds with well-developed seed coats, which have matured on the same plant, germinate much later.

The main question is whether the heteroblasty, which is of great ecological importance for survival, provides a general biochemical pathway, which, at the proper stage of maturation, affects seed germinability. Another possibility might be that in different species there are different biochemical pathways which are affected both by maternal position and by environmental factors. Whichever possibility is the correct one, the biochemical events involved in these phenomena are still not known. One can only speculate, as has been done in the past (Gutterman 1969, 1973, 1980/81a, 1982a, 1985) that, during seed maturation, different factors affect the accumulation of different amounts of materials which are involved later in the germination process of the seeds. They may react through three main pathways: (1) They could lead to the development of seed coats with different degrees of impermeability depending on day length and age effects; (2) These materials could result in an accumulation of germination inhibitors in the fruit, seeds and seed coats; (3) Some of these materials could be enzymes and hormones, such as ethylene.

Different amounts of ethylene are released from maturing tomatoes, depending upon the length of even a single night, and an additional amount of ethylene increases the level of ethylene release (Table 27). In *Cucumis sativus*, an additional quantity of ethylene changes to the opposite direction the levels of germination of seeds matured under SD or LD (Table 28). It is possible that enzymes and other materials that accumulate in the seed in different quantities, depending on environmental conditions during seed maturation, could affect seed germinability. This might explain the different levels of germination found in seeds of *Portulaca oleracea* when, after various treatments during maturation, these are transferred from imbibition at low to imbibition at high temperatures (Table 34) (Gutterman 1974). Achenes of *Lactuca serriola* (Fig. 39) (Gutterman 1992c) and some species of the Aizoaceae (Fig. 43 – 45) (Gutterman 1991), harvested at the same time from the same plants and in the same conditions, reached different levels of germination at varying temperatures.

Day length during the last stage of seed maturation affects the permeability of the seed coat to water in some species of the Fabaceae. It has also been shown that day length effects are transferred from the leaves and affect the development of the seed coat in covered fruit. So far, neither the biochemistry of this process nor the material or materials which are transferred from the leaves to the seeds and thereby affect the degree of seed coat development have yet been identified (Tables 23, 26; Fig. 32).

In some plant species, the effect of day length on flowering differs from its effect on seed germination. It is, therefore, possible that two different biochemical pathways are involved in the regulation of these two mechanisms. For instance, in *Ononis sicula* and *Trigonella arabica*, LD accelerates flowering,

76 Phenotypic Effects on Seeds During Development

but SD increases the percentage seed germination. In some plants whose flowering is not affected by day length, the germination of the seeds is accelerated by SD. The tomato and *Cucumis sativus* afford examples. Seed germination is accelerated by LD in *Carrichtera annua*. (Sects. 2.4.2.2 and 2.4.3.)

Genotypic inheritance increases the fitness of a species to its natural habitat so that its seeds germinate at the right time and in the right place. Phenotypic influences result in an increase in the diversity which ensures the germination of only a portion of the seed population in one season in the right place and at the right time. Other seeds remain in the seed bank and germinate in the following season or seasons.

In at least some of the plant species inhabiting most extreme and unpredictable deserts, heteroblastic responses of seeds to maternal and environmental conditions contribute very strongly to survival. Catastrophes that would otherwise be caused by mass germination after heavy rain followed by a long dry period which might cause all the seedling to die are thus prevented, as are intra- and inter-specific competition (Sect. 6.2.1). In other desert species different mechanisms regulate dispersal and germination. Such mechanisms include seed dispersal by rain which has been studied in *Blepharis* spp. (Sect. 3.3.3.4). In these plants, the number of seeds that are released is regulated by dry hydrochastic and cohesic tissues in such a way that only a part of the seed bank stored and protected by the dead mother plant will be released and germinate in a single rain event. The same phenomenon has been studied in *Asteriscus pygmaeus*, but dispersal is caused by a completely different mechanism, as in some other plants of the group that disperse their seeds by rain (Gutterman 1990b; Gutterman and Ginott in press) (Sect. 3.3.3.2).

The position on the mother plant and environmental factors affect seed germination in three ways. Three questions, therefore, arise:
1. Which material or materials are transferred from the leaves to the seeds so that they affect germinability in *Trigonella arabica* (Table 23; Fig. 31) (Gutterman 1978a), in tomatoes (Table 26) (Gutterman 1978b) and in *Carrichtera annua* (Fig. 32) (Gutterman 1978c), and what are the physiological processes involved?
2. What physiological processes and position effects are involved in the accumulation of these materials, which include germination accelerators or inhibitors that affect seed germinability? Are there interactions between the different positions of seeds on a plant, on an inflorescence or in a fruit? The substance leached from the aerial propagules of *Emex spinosa* contains germination inhibitors, but that leached from the subterranean propagules does not. When soaked in water, a higher proportion germinates of the aerials (terminals) than of the subterraneans (Table 6) (Evenari et al. 1977). The terminals of the multi-seeded dispersal units of *Aegilops geniculata* germinate worst (Fig. 18; Tables 9, 16) (Datta et al. 1970, 1972a, b; Gutterman 1992a) while those in *Pteranthus dichotomus* germinate best (Fig. 17; Table 8) (Evenari et al. 1982). In the capsule (fruit) of *Mesembryanthemum nodiflorum* the terminal seeds germinate best, while those from the lowest group germinate worst (Table 17) (Gutterman 1980/81a).

Conclusions

3. What causes different relative levels of seed germination at different temperatures in *Portulaca oleracea* (Table 34) (Gutterman 1974), in *Cheiridopsis* spp., *Juttadinteria proximus* (Figs. 43–45) (Gutterman 1992a) and in *Lactuca serriola* achenes (Fig. 39) (Gutterman 1992c)? Seeds of *P. oleracea*, which have matured under different environmental conditions, were transferred during imbibition from low to high temperatures. At each temperature, the seed population reached a higher percentage of germination but at relatively different levels (Table 34) (Gutterman 1974). In *L. serriola* (Fig. 39) (Gutterman 1992c) and some species of Aizoaceae (Figs. 43–45) (Gutterman 1991), seeds harvested from the same plant, at the same time and from the same environmental conditions, reached different levels of germination at different temperatures. Seeds harvested from the same plants at different times and germinated at the same temperature also reached different levels of germination.

Despite the fact that the biochemistry of the maternal position and environmental factors affecting seed germination are not known, their ecological importance is obvious. One mother plant produces various seeds that differ in their germinability. Seeds also differ in their germinability even on one branch and according to the position of the seeds in one fruit or dispersal unit. In different plant species these phenomena act through different mechanisms, but the result is that only a portion of the seed population germinates after one rain event. This is also dependent on the history of each seed from the time of maturation to the time of germination (Chaps. 3, 4, 5). This is a very important survival mechanism, especially for plants inhabiting extreme deserts in which the amounts and distribution of rainfall are unpredictable.

3 Seed Dispersal and Seed Predation of Plant Species in the Negev Desert

3.1 The Annual Cycle of Seed Maturation and Dispersal of Some of the Common Plant Species in the Negev Desert Highlands

The importance of dispersal mechanisms is that they bring at least some seeds to the most propitious place for germination and seedling development. The great majority of annual plants inhabiting the Negev desert have very small seeds which mature and disperse at the beginning of the summer (Figs. 52, 53). Annuals with small seeds complete their life cycles in a very short time (Fig. 135; Table 65). Some plants produce very large numbers of seeds per 1 m^2. In 1972/73, a season with only 48 mm of rain, 13 plants per 1 m^2 of *Schismus arabicus*, which grows to a few cm in height (Fig. 54), produced 1800 seeds (grains). One year later, in 1973/74, after 155 mm of rain (about 100 mm is the average rainfall) 89 plants per 1 m^2 produced 10,000 seeds (0.5 × 0.7 mm; weight: 0.07 mg) in the same location (Table 63). *Spergularia diandra* (Fig. 55) plants produce even lighter seeds (0.018 mg). In 1973/74, 36 plants per 1 m^2 produced 32,000 seeds. *Filago desertorum* (Fig. 56), in the centre of a depres-

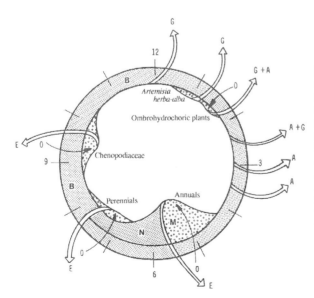

Fig. 52. Hypothetical drawing of seed dynamics in the soil showing the annual cycle of seed maturation, seed dispersal, and factors that reduce the number of seeds. Seed bank in the soil (*B*); accumulation of new matured seeds (*M*); additional amount added to seed bank (*N*); germination (*G*); aging and rotting seeds (*A*); seeds lost from the observation area, dispersed or eaten and/or collected by animals (*E*); seeds from outside the observation area (*O*). Months of the year (*3,6,9,12*)

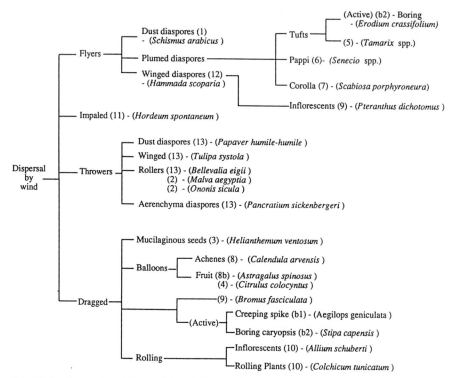

Fig. 53. A scheme of seed dispersal by wind of species inhabiting the Negev desert; the group No. (); and an example of one type

sion, produced as many as 36,000 seeds (achenes), and *Carrichtera annua* (Figs. 57, 58), produced 5,200 much larger seeds per 1 m^2 (Figs. 131, 132) (Loria and Noy-Meir 1979/80). Such large numbers of small seeds, as well as their long viability, are a result of the selective pressure of seed predation under unpredictable amounts of rain, and the short growing season (Sects. 6.1.2, 6.1.3).

This very important survival mechanism is concerned with escape from seed-eating animals and insects. This is especially important for annual plants which renew their populations from one year to another from seeds. In deserts where about 70% or more of the mature seed population is consumed each year mainly by ants, such as in *Scorzonera papposa* (Figs. 59–61; see below) the production of a large number of small seeds is possibly one of the strategies of escape (Sects. 6.1.2, 6.1.3).

The great majority of the seeds of the plant species on the Negev highlands mature at the beginning of, or during the summer when the activity of seed eaters and seed collectors is at its highest. Only a minority of plant species disperse their seeds in winter. Seeds of a few species are not dispersed at all and germinate in situ.

Plant species can be divided into five main groups according to the time of maturation, time of dispersal and the dispersal strategies: (1) maturation and

Fig. 54. *Schismus arabicus* plant in seed maturation, in its natural habitat on a flat loess plain, in summer. (Plant height: about 10 cm)

Fig. 55. *Spergularia diandra* dry plant after seed maturation in its natural habitat of flat loess soil. (Size: 10 cm)

dispersal at the beginning or during summer; (2) maturation at the beginning of summer and dispersal by rain in winter; (3) seed maturation in summer and germination in situ in winter; (4) maturation and dispersal in winter; (5) heterocarpy dispersal and species survival.

The yearly fluctuations in seed dispersal, and seed accumulation in the seed bank, are summarized in Fig. 52.

3.2 Seed Maturation and Dispersal in Summer

In the majority of plant species in the Negev – annuals, hemicryptophytes, such as *Scorzonera papposa* (Figs. 59, 60) and most of the geophytes, such as *Tulipa systola* (Fig. 62) – the seeds mature and are dispersed at the beginning of the summer (May/June). The seeds of perennials such as *Zygophyllum dumosum* (Fig. 63) and species of *Helianthemum* mature during July and August, while those of some of the common annual and perennial Chenopodiaceae mature during July to October, or even November. The seeds mainly dispersed during summer are those that are distributed by wind: dust seeds, balloon, winged, etc. (Figs. 53, 62–67). A minority of species have their seeds dispersed by ants, birds, mammals and other animals (Figs. 68, 75, 76). During summer, there are large fluctuations in seed numbers (Fig. 52).

Fig. 56. *Filago desertorum* plant showing capitula at the end of seed dispersal at the beginning of summer, in its natural habitat on the flat loess plain. (Size: 2–3 cm)

Fig. 57. *Carrichtera annua* dry dead plant with pods containing maturing seeds in its natural habitat on a flat loess plain during summer. (Size: 20 cm)

3.2.1 Seed Dispersal by Wind

Seeds that are dispersed by wind can be divided into at least 14 clearly defined groups (Fig. 53):

3.2.1.1 Passive Dispersal by Wind

In areas, such as flat loess plains, where the bare soil is covered with a massive crust, there are two main micro-habitats in which seeds accumulate and ger-

Seed Maturation and Dispersal in Summer

Fig. 58. *Carrichtera annua* dry pods after seed maturation, on the dead mother plant (×2)

Fig. 59. *Scorzonera papposa* flower bud and open capitulum before achene dispersal (×1)

Fig. 60. *Scorzonera papposa* capitulum with achenes with open pappi in the capitulum

Fig. 61. An ant nest surrounded with remains of *Scorzonera papposa* achenes. *On the left S. papposa* plant from which achenes have already been dispersed from two of the three inflorescences

minate: soil cracks and furrows (Fig. 113a, b). After having been shed, seeds are carried to such places by the wind. They may be ejected from the plant onto the bare soil, or entangled on the dead mother plant (Fig. 53). Run-off water in such places accumulates giving the seedlings a good chance of survival.

Plant species with tiny, dust-like seeds as an "escape strategy" preventing mass consumption. The most typical means of dispersal by wind is through the production of numerous dust-like seeds. These are found in *Schismus arabicus* (Poaceae) (Fig. 54), *Spergularia diandra* (Caryophylloceae) (Fig. 55), *Nasturtiopsis coronopifolia*, *Arabidopsis kneuckeri*, *A. pumila*, *Diplotaxis harra*, *D. acris* (Cruciferae), and many other species. Such tiny seeds are carried by the wind to cracks or furrows in the soil and are covered by dust. *S. arabicus* is one of the most common annual grasses producing a tremendous amount of dust-like seeds, 0.5 – 0.7 mm in size and weighing about 0.07 mg (Loria and Noy-Meir 1979/80). In their size and colour they resemble particles of sand. Some of them are harvested by ants, together with the whole inflorescences, before full maturation and dispersal.

Roller diaspores are seeds with hard and round seed coats, 1 – 2 mm in diameter. Examples are provided by the seeds of *Malva aegyptia* (Malvaceae) or *Bellevalia desertorum* and *B. eigii* (Liliaceae) (Fig. 64), which are dispersed by gusts of wind, and *Ononis sicula* (Fabaceae), whose seeds are dispersed by

Fig. 62. *Tulipa systola* with corolla surrounding seeds. Capsules are opened along the three valves and seeds are released by blasts of wind. (Size: 20 cm)

Fig. 63. *Zygophyllum dumosum* winged diaspores shortly before maturation

explosion of the pod. The seeds of these species, and of many others, are blown across the bare soil surface by the wind, and deposited in cracks or depressions of the soil. Here they, too, become covered with dust, which may protect them from seed-eating animals (Evenari and Gutterman 1976; Loria and Noy-Meir 1979/80; Gutterman 1982a). In mid-October 1991, after the first rain, some seeds of *B. desertorum* still remained in the capsules of inflorescences which were lying on the soil surface (Sect. 3.3.1).

Small seeds covered with mucilage (myxospermy – Sect. 3.3.1). These seeds are rolled by the wind across the desert and then, after wetting by rain or dew, adhere to the soil crust by mucilage (e.g. *Helianthemum ventosum* and *H. vesicarium*).

Unopened fruit as dispersal units (synaptospermy – Sect. 3.3.1) (Murbeck 19/1920). These too are rolled by the wind when the fruit becomes dry, lightweight and hollow (e.g. *Citrullus colocynthis* (L.) Schrad., *Astragalus spinosus*).

Fig. 64. *Bellevalia eigii* in seed dispersal. Seeds at the bottom of the capsule are dispersed later in the summer

Seeds with tufts of hair, e.g. *Caralluma europaea, Calotropis procera* (Fig. 65), *Gomphocarpus sinaicus* (Asclepiadaceae), *Tamarix* spp., *Reaumuria negevensis* and *R. hirtella* (Tamaricaceae) as well as *Trachomitum venetum* (Apocynanceae). Seeds of *Tamarix* which enter free water in wadis and pools during summer, germinate where the water layer is about 1–3 cm deep. *Hairy nutlets* are dispersed by wind, e.g. a desert ecotype of the Mediterranean plant species *Anemone coronaria* L.

Seeds (achenes) with a pappus of hairs or bristles, which may be branched (Fahn 1967), are a common means of dispersal among the Asteraceae. Ninetyseven of the 250 species of Asteraceae in Israel inhabit the deserts. Fortysix of those inhabiting the deserts of Israel have achenes with a persistant pappus, 26 are dimorphic and 25 have a deciduous pappus which reduces long-distance dispersal (Feinbrun-Dotan 1978; Shmida 1984). This gives an advantage to seeds in favourable depressions to germinate in the same locations as their mother plant (Ellner and Shmida 1981). A negative correlation was found (Shmida 1984) between heterocarpic species and species with deciduous pappi. Shmida (1984) hypothesized that heterocarpy is an alternative to the character of the deciduous pappus in species of the Asteraceae.

Examples of achenes with a pappus are provided by *Scorzonera papposa* (Figs. 59, 60), *S. Judaica, Senecio* spp. and many other species whose seeds can

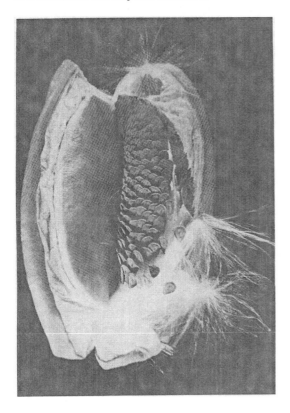

Fig. 65. *Calotropis procera*. A hygrochastic opening of the endocarp during the day and seed dispersal by wind in the afternoon. (Size: 10 cm)

be dispersed over very long distances by the wind, or are blown across the bare soil surface into furrows and depressions. In Saharo-Arabian species, such as *Ifloga spicata* and *I. rueppellii*, the achenes of the central staminate florets have a pappus while those of the peripheral pistillate florets lack a pappus. This has also been found, for example, in the Saharo-Arabian and western Irano-Turanian *Filago desertorum* (Fig. 56) (Wagenitz 1969; Feinbrun-Dothan 1970).

Several species of *Crepis*, *Picris* and *Geropogon* have central achenes with pappi (which are easily detached when ripe) and peripheral achenes (with rudimentary pappi). These are securely connected to the receptacle and the peripheral achenes are not discharged until the winter, when they fall near the plant (Danin 1983).

In the dimorphic achenes of *Heterotheca latifolia*, distributed from northern Mexico to New York, differences in dispersal were found between the long-dispersal disc achenes with a pappus and the short-dispersal ray achenes lacking a pappus. Both types develop on the same capitulum but have different germination in time and space (Venable and Levin 1985).

Fig. 66. *Scabiosa porphyroneura* with fully developed corolla, at the beginning of dispersal

Seeds with a corolla, are found in *Scabiosa porphyroneura* Blakelock (Dipsacaceae), a West Irano-Turanian and East Sahara Arabian annual plant (Fig. 66).

Seeds with coats formed by large air-containing cells. Pancratium maritimum inhabits the Mediterranean region, extending along sandy shores of the Atlantic, Black and Caspian Seas (Feinbrun-Dothan 1986). In Israel it inhabits the first vegetation belt of the Mediterranean seashore. Along this sandy shore, plants such as *Artemisia monosperma* and *Retama raetam* of the Saharo-Arabian phytogeographic regions are found. The seeds of *Pancratium maritimum* mature in autumn and their coats are formed by large air-containing cells. Some seeds can be carried by the autumn eastern wind to the sea. They are able to float on the surface of the sea for more than 17 days without losing their ability to germinate. After 17 days of floating, seeds germinated to 78% in distilled water, in the dark, after washing. Furthermore, seeds which had sunk below the sea surface for more than 40 days still germinated to 48% in the dark, after washing. This was reduced to 15% after 50 days and 3% after 60 days. Seeds are thrown back onto the shore during westerly winter storms. In order to germinate under natural conditions, they have to be washed by rain and buried in the sand because salinity and light inhibit their germination (Keren and Evenari 1974).

Fig. 67. *Calendula arvensis* with cymbiform achenes before dispersal, in addition to two other types of achenes: one is atelechoric and small and is dispersed near the dead mother plant; the other is larger with protrusions that adhere to animals and are, thus, dispersed to further distances (exozoochory)

Other plants which inhabit the deserts of Israel have similar seed coats and are dispersed by wind. *Pancratium sickenbergeri* is found in the sands of the west, north and central Negev, Dead Sea and Arava Valley extending into the Saharo-Arabian region. Similar seed dispersal is found in plants such as *Urginea maritima*, inhabiting the Mediterranean region, Judean desert and central Negev, *U. undulata*, inhabiting the West-Irano-Turanean region, the Judean Desert, Central Negev and Dead Sea area, and *Scilla hanburyi*, which inhabits the Saharo-Arabian region, extending into the Mediterranean region (Feinbrun-Dothan 1986).

Cymbiform achenes (balloons) are exemplified by *Calendula arvensis* (Asteraceae) (Heyn et al. 1974) (Fig. 67). These roll along the soil surface and fall into cracks or accumulate in large numbers in depressions.

Balloon-like fruit. The fruit of *Astragalus spinosus* are balloon-like.

Passive movement of multi-seeded dispersal units. Relatively large dispersal units contain from one to several seeds. Examples are afforded by the inflorescence of *Pteranthus dichotomus* (Caryophyllaceae) (Fig. 53), or the pod of *Trigonella arabica* (Fabaceae) (Fig. 31). The dispersal units of other species, such as the spike of *Bromus fasciculatus* (Gramineae), contain about three spikelets. The lower ones have one caryopsis 8 mm long and 0.5 – 1 mm in diameter. The dispersal units are moved passively on the soil surface by the wind and accumulate in depressions, cracks, or between stones.

Rolling plants or inflorescences. Stems are dehiscent near the soil surface. The disconnected part of the plant (only the inflorescence or the inflorescence with the leathery leaves), is carried by the wind and rolls along like a wheel (e.g. *Allium rothii*, *A. schuberti*, *Gundelia tournefortii*, *Colchicum tunicatum*, *C. ritchii*). Infloresences of *A. rothii* were found still to have some of their seeds trapped in the dry capsules after the first rains of the season, in mid-October 1991. At this time, relatively few of the tremendous number of seeds that had been shed in a patch of *A. rothii* during the summer were still to be found on the soil surface (10 per 1 cm^2 in a patch where there is a crowded population of this plant). Because of the low percentage of seed synaptospermy (only ca. 5% of the seeds remaining in the inflorescences), this is very important as the seeds remain protected until the time of germination. These rolling inflorescences are trapped in porcupine diggings and other depressions, between stones and on hillocks beneath shrubs which are all relatively better microhabitats for seed germination. In some depressions of 100−150 cm^2 90−160 seeds were counted near Sede Boker.

Entangled dispersal units are found in *Hordeum spontaneum*. The grains are even larger than those of *Aegilops geniculata*, and are well protected both by the unit and by the system of dispersal. In May and June, this winter annual disperses a single spikelet containing one grain, from its multi-spikelet spikes. The dispersal units fall and are entangled in the dry remains of the dead mother plant. The wind shakes the long (12−15 cm) awns. This mechanism acts like a bore, because the direction of the bristles enables the dispersal unit to move in one direction only until it reaches a depth of 2−3 cm below the soil surface. Here it remains until it germinates in the rains of the following winter or winters. This system protects the caryopsis in a very efficient way. Almost the same number of dispersal units with grains per unit area was found in a patch inhabited by *H. spontaneum* near Sede Boker in 1988 at the beginning of the summer and before the first rain at the end of the summer.

Winged diaspores, such as those of *Salsola inermis*, *S. volkensii*, *Hammada scoparia* (Chenopodiaceae) and *Zygophyllum dumosum* (Zygophyllaceae) (Fig. 63), accumulate between stones or beneath the mother plant and germinate there during the following growing season.

Ballist seeds. These are produced by plant species with long florescent stems which terminate in a capsule or capsules and open along the dehiscence zones when the seeds mature. The force of the wind causes separation of the seeds from the fruit in *Papaver humile* Fedde. subsp. *humile*, in which the dust-like diaspores are shaken from the capsule by the wind. The same occurs in *Ixiolirion tataricum*. There are also species whose seeds are surrounded by a corolla as in *Tulipa polychroma* and *T. systola* (Fig. 62). After separation, the seeds are carried by the wind. It was observed in the Negev highlands near Sede Boker, in mid-October 1991 that more than 40% of the open dry capsules of *T. systola* as well as *T. polychroma* which remained in the area, still contained seeds. This was just a few days after the first rainfall of the 1991/92 rainy season. In ca. 10% of these capsules, nearly half of the seeds remained in the lower part of the open, dry capsule where they had matured. In a few capsules situated in the centre of *Artemisia sieberi* (small shrubs) nearly all the seeds were still in the capsules (Kamenetsky and Gutterman, in press). (The importance of these observations is mentioned in Sect. 3.2.3). In *Urginia undulata*, *U. maritima*, *Pancratium sickenbergeri*, *Scilla hanburyi*, etc., the seed coat is formed by large air-containing cells, like a sponge. The seeds are carried either by wind or flood (Sect. 3.2.1.1). In *Iris petrana* the elaiosome (food body) seeds are rolled along the soil surface by the wind after dispersal and collected by ants (Fig. 68) (Beattie 1985).

Fig. 68. *Iris petrana* seeds (size: 3 mm) with arils that attract ants which collect them. The seeds are thrown from the open capsules by the action of the wind

3.2.1.2 Active Dispersal

'Creeping' dispersal units. Aegilops geniculata (= *A. ovata*) has dispersal units with two to four spikelets (Fig. 18) and three to six relatively large caryopses (Table 16) (Sect. 2.2.3). The dispersal units of a Jerusalem population, extending into the Judean desert, have hygrochastic awns which open within 10–20 min after wetting, and close within 2–3 h of being dry (Fig. 69) (Gutterman, unpubl.). This hygrochastic movement causes a 'creeping' motion (Ulbrich 1928) after wetting by dew during the nights and drying again during the following days. This is in addition to passive movement by wind on the soil surface. The dispersal units bore by means of the awns above the ground as they are shaken by the wind (see *Hordeum spontaneum* – Sect. 3.2.1.1). Other creeping species are: *Avena sterilis, A. wiestii, A. barbata,* in which the lower part of each of the two awns twist when dry and straighten when wet. In all these, there are two means of movement: passive movement by wind and hygrochastic movement. The awns and other parts of the dispersal units are constructed so that the bristles, which are pointing in the distal direction, cause the dispersal unit to creep in that direction only.

Boring dispersal units (Trypanocarpy) (Zohary 1962). *Stipa capensis*, one of the common annual grasses of the Negev, has dispersal units of one caryopsis spikelet with one awn of which the lower part twists when dry and straightens when wet. In addition to passive movement caused by the wind, this mechanism acts like a borer which brings the caryopsis 2–3 cm below the soil surface, the place at which it germinates during the following winter or winters (Gutterman 1982a). The swivelling motion is assisted by a combination of dew at night and low humidity during the day. It occurs mainly during summer after the time of dispersal. During summer more than 90 nights of dew have

Fig. 69. *Aegilops geniculata* (= *A. ovata*) collected from a population in Jerusalem. The hygrochastic awns of the spikelets, which are the dispersal unit, open by wetting within 10–20 min and closes when dry within 2–3 h. This hygrochastic movement causes a creeping motion of the spikelet. (Size: 3 cm) (Fig. 18)

Seed Maturation and Dispersal in Summer

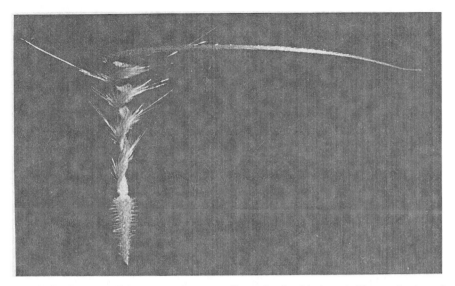

Fig. 70. *Erodium crassifolium*. A mericarp as a dispersal unit with the swivelling mechanism of the lower part of the 'beak', when dry. (Size: 10 cm)

been measured at Sede Boker. Almost the same mechanism is found in *Erodium oxyrhynchum* and *E. crassifolium* (Geraniaceae), after the hairy dispersal units have been transported by the wind (Fig. 70) (Fahn 1967; Zangvil and Druian 1983) (Sects. 1.2.2, 4.1.3.3).

Seeds dehisced by exploding pods by the twisting of the valves after maturation (Overbeck 1925; von Guttenberg 1926; Fahn and Zohary 1955; Fahn and Werker 1972). The round seeds of *Ononis sicula* dehisce explosively from the pods. They are then rolled by the wind on the soil surface until they accumulate in depressions, cracks or between stones (Fig. 24).

3.2.2 Porcupine Diggings and Other Depressions as Wind-Traps for Seeds, and the 'Treasure Effect'

Many seeds accumulate between stones or in cracks, as mentioned above, but one of the most important micro-habitats in the Negev highlands and some other deserts of Israel, are porcupine (*Hystrix indica*) diggings (Fig. 113 a, b; Sect. 4.1.2). These form a wind trap for many of the seeds that are blown by the wind across the bare soil surface (Fig. 71). Organic matter and runoff water accumulate in porcupine diggings and create a favourable micro-habitat. These diggings are scattered, sometimes at a density of more than one furrow per 1 m^2 (Gutterman and Herre 1981; Gutterman 1982c). Similar activity of *Hystrix africaeaustralis* has been found in the Namaqualand desert of South

Africa. At least one type of geophyte, such as *Homeria shlechteri*, had been totally consumed in great numbers and subterranean roots and stems of one succulent shrub species, such as *Herrea elongata*, had been partially consumed in great numbers and renewed themselves in the diggings. The patches where these species were consumed were found to contain diggings at a density of even more than one digging per 1 m^2 (Gutterman, original data). Seed distribution is correlated with the dimensions of micro-topographic depressions, as was also observed in the Sonoran desert (Reichman 1984). In some places in the Negev there are patches where one to three diggings per m^2 have been observed. Porcupines dig the soil to obtain corms, bulbs, tubers, and the underground parts of leaves and stems, which they eat. Eighteen species of geophytes and hemicryptophytes have been found to be consumed by porcupines in the Negev Desert highlands. The importance of porcupine diggings for renewal of the vegetation, both by germination and by vegetative propagation, is obvious (Gutterman 1982c, 1987, 1988a). The initial depth of the diggings depends upon the depth of those parts of the plant that the porcupine is gathering. Diggings range from 5–30 cm in depth and are about 30 cm long and from 10 to 15 cm in width (Fig. 113a, b). In such micro-habitats, up to ten times more of the seedlings of annual plants survive than in surrounding areas. As the percentage of cover-over of the digging increases over the years, a succession of annuals has been observed, as well as changes in the number of

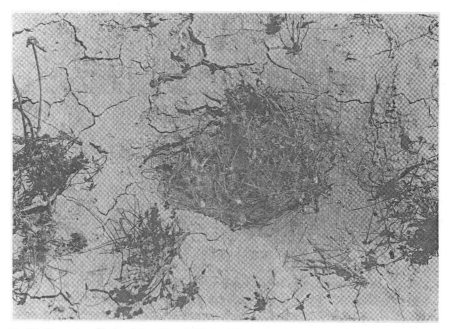

Fig. 71. Porcupine digging (size: 20 cm) filled with dispersal units, seeds, etc. which have accumulated by means of wind during summer

Seed Maturation and Dispersal in Summer

plants, the variety of species and total biomass (Gutterman et al. 1990). Seed yield may be 16 times greater (Sect. 6.3; Figs. 137, 138) (Gutterman 1989a).

In contrast to the enormous numbers of seeds that sometimes accumulate as a result of the wind during summer, and also of runoff water in winter (Gutterman 1989a; Gutterman et al. 1990) (Fig. 71), there is not always a massive germination. This is because of the 'treasure effect' (Gutterman 1988a). As more seeds accumulate in a small area, the probability that a seed eater will arrive and remove nearly all of them is much greater than when only a small quantity of seeds accumulate. This 'treasure effect' has also been studied in larger depressions. When plots (25×25 cm) were irrigated with different amounts of water, they produced different yields of the seeds of *Malva aegyptia*. In the area that received most water (the equivalent of 150 mm), about 3700 metacarps ('seeds') were counted per 0.25 m^2 but in squares that received less water (the equivalent of 90 mm) the yield of seeds was only 1600 'seeds'. The seeds that had been dispersed on the soil surface at the beginning of summer were collected by seed eaters during the summer. In squares that contained the largest numbers of seeds at the beginning of the summer, only a few seeds remained at the onset of the following winter. In squares that contained less seeds, the majority were still present before the growing season. Thus, paradoxically, in the following winter far less seedlings appeared in the places which had contained the greatest amount of seeds (Gutterman 1982a).

3.2.3 Seed Predation and Seed Dispersal by Ants, Birds and Mammals

More than 70% of the seed production is gathered and eaten by granivores such as mammals, birds and insects (mainly ants) in some years in certain desert areas (Evenari et al. 1971; Gutterman 1982b; Abramsky 1983; Bar et al. 1984). This applies to the Negev desert and to the arid regions of Australia, as well as North and South America (Morton 1985). In mesquite scrub habitats of Arizona, which are virtually a monoculture of the perennial woody shrub *Prosopsis juliflora*, little seed predation has been observed either by rodents or ants. This was not the case in two mixed desert shrub habitats, 150 km apart. There, both rodents and ants removed large numbers of seeds. The rodents were found to be much more efficient than the ants (Brown et al. 1975). This was also observed to an even greater extent in the Sonoran Desert (Reichman 1979). Ants take seeds from the surface, but seeds buried more than 1.5 cm are not available to them (Tevis 1958a; Bernstein 1974). No foraging, either by ants or rodents, occurs in bushes (Reichman 1975). The same has been found in some synaptospermic geophytes of the Negev (Kamenetsky and Gutterman, in press). This is not the case in the Negev Desert highlands, according to Hord (1986), in other plants. Harvester ants (*Messor arenarius*) have been seen to cut stems of *Reboudia pinnata* with two to three pods, stems of *Salvia lanigera* with seven fruits, a capitulum of *Senecio glaucus*, flower buds of *Thymelaea hirsuta*, inflorescences of *Plantago coronopus* and pods of *Ononis sicula*. *M. eveninus* also cut through a stem with a pod of

Reboudia pinnata, as well as fruits of *Helianthemum ventosum* and *H. kahiricum*.

An obvious correlation exists between seed colour and soil colour. This has been noted in particular habitats of many of the 19 local populations of *Salvia columbariae* from the southern Californian deserts to the coastal valleys of California (Capon and Brecht 1970). The colour of the grey-brown or red-brown seeds closely matches the range of colour of soil in the local habitats of each ecotype. Seed coats are patterned to resemble soil particles. On grey soil only grey seeds remained of two groups of 1000 of each of the two colours at a site where ant activity had not been seen. Seeds having the most cryptic colours survive in places where birds feed intensively, but only few survive in areas from which ants collect seeds (Brayton and Capon 1980). On the other hand, studies on the fate of seed dispersal by ants (myrmecochory) (Warburg 1892) show that seeds that were not collected by ants were eaten within a few hours by other animals (O'Dowd and Hay 1980; Beattie 1985).

In the forests of West Virginia, ants are essential to the survival of seeds of some species and, later on, to the appearance of seedlings (Heithaus 1981). Nests of harvester ants are a primary habitat for a few ruderal plants such as *Silybum marianum* in Israel. The achenes have an oily food body which attracts the harvester ants. These collect the achenes and bring them to their nests. The oily bodies are removed, and the intact achenes carried to the refuse zone of the nest where they germinate. The soil near the nest is much more fertile than that of the surrounding area. The plants that grow there are larger, and each produces three to four times more heads. Other plant species that benefit from this special habitat in the Mediterranean and Irano-Turanian regions of Israel include *Malva parviflora*, *Beta vulgaris*, *Chrysanthemum coronarium* and *Echium judaeum* (Danin 1989). Of 39 plant species in the natural grazing areas of the northern Negev, only 11 are included in the total of 15 that grow on the hills of ants (*Messor* spp.). These are species known to prefer humidity, aeration and nitrogen (Ofer 1982).

Near Sede Boker *Reboudia pinnata* benefits from the refuse zone of *Messor eveninus*. The valves of the pods of *R. pinnata* separate after 1.5 h of wetting by rain drops, and the seeds are dispersed (Sect. 3.3.2.2). The unopened terminal part of the pod (the beak) contains a single seed. This falls to the soil surface from where it may be dispersed by wind or collected by ants and carried to their nest. Many of these seeds germinate from the 'beaks' subsequently discarded into the nest's refuse zone. This has a radius of about 1 − 1.5 m from the centre of the nest. In spring, the existence of a nest can be deduced from a distance by a group of crowded plants about 40 − 50 cm tall. Other seeds that are dispersed by rain from the dead plants of the previous year adhere to the soil surface and germinate each winter in the nest's refuse zone near the dead mother plant (original data).

Near Phoenix, Arizona, in the Sonoran Desert, 6 of 36 plant species surveyed were associated with ant nests. They grew within 1 m of nest entrances of *Veromessor pergondei* or 1.5 m from those of *Pogonomyrmex rugosus*. *Schismus arabicus* benefits from growing on refuse piles by a 15.6-fold increase

Fig. 72. *Plantago coronopus* in its natural habitat on the flat loess plain or on slight slopes, in summer. The leaves are dry and the mature seeds are enclosed in the dry inflorescences. (Size: 5 cm)

in the number of grains/plant and 23.7 of total grains mass/plant, in comparison with a control (Rissing 1986). *S. arabicus* is very common in the Negev Desert but there is no evidence of any such connection with ant nests.

In summer, when food is short, harvesting ants separate the dry woody axis of *Plantago coronopus* and carry the whole inflorescence, containing seeds, to their nest (Figs. 72–74). In the case of some plant species, such as *Schismus arabicus* (Fig. 54), ants harvest the whole inflorescence. Whole pods of *Ononis sicula* are harvested before seed dispersal at the beginning of summer (see above, Hord 1986).

In the Negev, ants collect the majority of seeds harvested each year by various kinds of animals. In the Northern Negev, west of Beer Sheva, an ant nest was experimentally supplied with wheat grains. The ants collected about 25 kg of wheat grains within a few hours (M. Loria, pers. comm.). Ants are likewise most important seed removers in the semi-arid Karoo Desert of South Africa: rodents remove less and birds the least number of seeds (Kerley 1991). In the Sonoran Desert of North America, rodents are primary consumers, ants are also very active, while birds were unimportant. In the Monte Desert of South America, ants are the main granivores and birds are not important (Mares and Rosenzweig 1978). In the Chihuahuan desert, ants are the major seed predators, then rodents (Davidson and Samson 1985).

An experiment was conducted from November 1984 until March/April 1985, near Sede Boker. In this 2400 seeds of *Tulipa systola* (Fig. 62), *Bellevalia*

Fig. 73. *Plantago coronopus* woody inflorescences: *on the left* dry; *on the right* opening and seed dispersal by rain when wet. (Size: 3–4 cm)

desertorum and *B. eigii* (Fig. 64) were spread, in groups of 25, on the surface of two hill slopes and in a wadi. They were surrounded by a circular plastic wall, 9 cm in diameter and 1.5 cm high. Nearly all the seeds had been removed after a few days and, by March/April, all the 2400 seeds had disappeared from all three habitats. Ant nests were situated all over the experimental site but seeds may also have been collected by other seed consumers including birds and mammals (Boeken 1986; Boeken and Gutterman 1990).

As mentioned in Sect. 3.2.2, it is possible that *Tulipa systola* and other synanthous geophytes may overcome the intensive collection of seeds from the soil surface by keeping a relatively large number in the dry open capsule during summer. The seeds are later dispersed by strong winds of the storms that bring rain. These blow the seeds out of the capsules and disperse them near the time suitable for germination (Kamenetsky and Gutterman, in press).

Fig. 74. *Plantago coronopus* mucilaginous seeds on the wet soil after dispersal

Retama raetam (Fabaceae) is a large evergreen shrub of the Saharo-Arabian and Irano-Turanian geographical region. It inhabits wadi beds and slopes, as well as stable and shifting sands on the Mediterranean coast (Zohary 1972). The pods, including the seeds, are consumed by Leporidae (*Lepus capensis* and *L. europeus*) and goats. The undamaged seeds are found in their faeces (Fig. 75). These seeds germinate much faster (50%) than seeds that have not been eaten. Of the latter, only 2% germinated experimentally (Charif, Y., pers. comm.).

Of three Acacia species (Mimosaceae), *Acacia raddiana*, *A. tortilis* and *A. gerrardii* ssp. (*negevensis*), inhabiting the Negev and Sinai Deserts, the pods are eaten by gazelles (*Gazella dorcas*) and 64–99% of seeds are infested by species of seed beetles (bruchids). Ibex (*Capra ibex nubiana*), Beduin goats and camels also feed on the Acacia pods. Seeds of *A. raddiana* with intact embryos that were fed to gazelles germinated to 21% after removal from the faeces, in comparison with 4% of the controls after ca. 10 days of imbibition. In this experiment 72% of the seeds were infested by seed beetles (Halevy 1974).

The nitre bush, *Nitraria billardieri* DC (syn. *N. Schoberi* L.) (Zygophyllaceae) inhabits overgrazed areas of bladder saltbush, *Atriplex vesicaria*, communities in the arid zone of Riverine Plain, Australia (Leigh and Noble 1972). Emus (*Dromaius novaehollandiae*) have been observed to consume the ripe fruits of the nitre bush (ca. 11×5 mm, weighing ca. 60 mg), and up to 1350 seeds per faecal dropping were recorded (Noble 1975a). The germination of the emu-ingested seeds was 50% after 4 days of imbibition, in comparison with only 3% of hand-collected seeds. Emus are most important in the successful

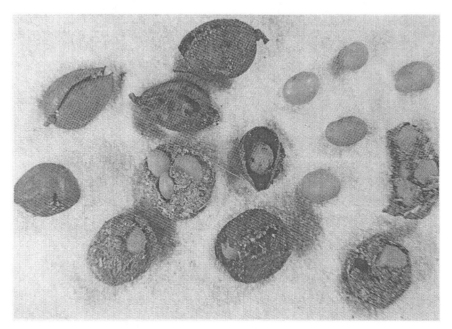

Fig. 75. *Retama raetam* seeds in pods free seeds (size: 0.6 cm), and seeds that had been consumed and released in time from the faeces of *Lepus* spp. and possibly other vegetarian animals that are found in sandy desert habitats

establishment of nitre bushes on heavy clay soils where seeds are rarely buried (Noble 1975b).

In *Sternbergia clusiana* (Amaryllidaceae) of the Negev, the pericarp of the fruit cracks when the fruits become weak at the time of seed ripening. Tension in the fruit is increased by the swelling of the arils (elaiosome) of the seeds and the shrinking of the pericarp. Seeds are collected by birds and ants from the splits that appear in the fruit (Fig. 76). In *Iris petrana*, the seeds are shaded with arils which are attractive to ants (Fig. 68). In *Loranthus acaciae* (Loranthaceae), the fruit is covered by a sticky substance so that it adheres to the beaks and legs of birds. Thus, it is carried from one tree to another and the seeds germinate on the branches. These seeds do not require water for germination. This was found by Galil (1938) to be the case in *Viscum cruciatum* (Loranthaceae), which germinates on the branches of olive trees.

3.2.4 Changes in the 'Seed Bank' During Summer

In the Negev Desert, with the exception of porcupine diggings (Fig. 113a, b, c), nearly all seeds are situated on the soil surface crust and to a depth of 20–25 mm (Gutterman et al. 1982). In Sonoran Desert soils, 89% of seeds are situated in the ground, to a depth of 20 mm (Childs and Goodall 1973). The

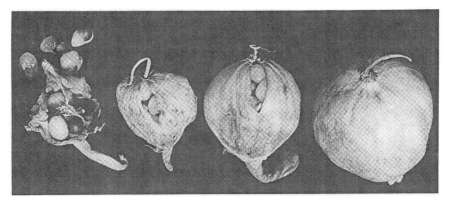

Fig. 76. *Sternbergia clusiana* fruit showing the cracking of the pericarp when the tension in the fruit is increased by the swelling of the arils of the seeds. *On the left* seeds dispersed from the dry fruit or collected by ants or birds (×1)

number of seeds in the soil is very variable, depending on micro-topographical depressions, and can be between 4000 and 63 800 per m^2 (Reichman 1984).

In a very preliminary experiment in June and October 1981, four soil samples were taken (Fig. 52): from 1 m^2 of soil crust at a depth of 0–0.5 mm near Sede Boker, from the three main habitats of the Negev: (a) Hammadetum (flat loess areas or wide wadis with a loess bed), (b) Artemisietum (on northern and western facing slopes) and (c) Zygophylletum (on southern and eastern facing slopes) (Table 36). Hammadetum is the main habitat of annual plants: conditions in the Zygophylletum are the most extreme. Seeds were separated from these soil samples. The number of seeds ranged from 3250 to 200, depending on the season and the habitat. From Table 36 it can be seen that there was a reduction of 29% in the amounts of seeds collected in Hammadetum in October, compared with July, in comparison with a 72% reduction in Zygophylletum. There are relatively few annuals in Zygophylletum and the pressure from seed eaters is the highest.

Table 36. Preliminary results calculated to show the number of seeds per square meter (1 m^2) in the soil crust (0–5 mm depth) of the three main habitats, at the beginning of the summer (June) and before the first rain (October). These calculations were made from seed separation of four soil samples of 120 g (100 ml) from each habitat collected in 1981. (After Gutterman et al. 1982)

Habitat	Number of seeds per m^2		% Seeds remaining in October in comparison with June
	Samples collected in		
	June	October	
Hammadetum	3243	2298	71
Artemisietum	990	1388	140
Zygophylletum	731	205	28

Table 37. The average number of seedlings per sample that appeared in pots after wetting in winter 1981/82 of soil samples (120 g) taken from the three main habitats in the Negev Desert highlands near Sede Boker. Soil samples from the soil surface, soil crust and soil depth from each of the three habitats were tested. Final observations made on 20 December 1981. (After Gutterman et al. 1982)

Habitat	Average number of seedlings per sample		
	Hammadetum	*Artemisietum*	*Zygophylletum*
Soil surface	42.6	32.6	13.6
Soil crust (0 – 5 mm)	13.6	11.0	9.3
Soil depth (5 – 20 mm)	8.7	7.0	8.2
Habitat average	21.6	16.9	10.4

There were relatively few annual plants in Artemisietum because of competition with *Artemisia sieberi*, the dominant perennial. The increase of ca. 40% in the number of seeds in October compared with the June samples is the result of species such as *Zygophyllum dumosum*, *Salsola inermis*, *Noea mucronata*, whose seeds mature and disperse during summer (Fig. 52) (Gutterman et al. 1982). This is in addition to the synaptospermic species in which there is a delay in seed dispersal after maturation (Sect. 3.2.5).

In order to find out how many seeds were ready for germination, loess soil samples from near Sede Boker were placed in a 5 mm layer on sterile sand in 10 cm diameter pots and moistened. In samples from Hammadetum, the maximum number of seedlings appeared in all the three layers tested (soil surface, crust 0 – 5 mm depth, and 5 – 20 mm). Zygophylletum samples showed a minimum number of seedlings except from a depth of 5 – 20 mm. The largest number of seedlings appeared in samples taken from the soil surface in all three habitats, while the lowest numbers appeared in samples from the deepest soil (Table 37). Seeds from the soil surface, where the majority ready for germination were found, were least protected. The seeds of species that have no mechanism to protect them are harvested in great numbers during summer by seed-eaters (Tables 36, 37, 38, 38 a) (Gutterman et al. 1982).

3.2.5 Seed Dynamics in the Soil of Species Dispersed by Wind

During the same experiment, seeds were identified from the samples (120 g) of soil from the three habitats. Out of a total of nine species with wind-dispersed seeds, 83% found in the June samples were also found in October. Of plants whose seeds are dispersed by wind, less were found in Zygophylletum, more in Hammadetum and many more in Artemisietum (Table 38 a). *Roemeria hybrida*, *Herniaria hirsuta* and *Spergularia diandra* (Fig. 55) (Zohary 1962) have a synaptospermic mechanism. There is a long delay in the dispersal of their

Seed Maturation and Dispersal in Summer

Table 38. Numbers of seeds of different plant species, whose seeds are dispersed by rain, separated and identified from 120 g samples of soil from the soil crust (0 – 0.5 mm) layer collected in June and October 1981 from the three habitats: *Hammadetum* (H.s.), *Artemisietum* (A.h.a.) and *Zygophylletum* (Z.d.). Seeds from two samples from each habitat were identified. (After Gutterman et al. 1982)

Season	Number of seeds in samples collected in June 1981			Number of seeds in samples collected in October 1981		
Habitat	H.s.	A.h.a.	Z.d.	H.s.	A.h.a.	Z.d.
Plant species						
Reboudia pinnata	–	8	10	–	8	2
Trigonella stellata	45	6	87	45	2	3
Astragalus tribuloides	–	1	26	2	2	–
Plantago coronopus	12	15	2	3	5	–
Aizoon hispanicum	1	1	–	–	–	–
Total	58	31	125	50	17	5
Percentage of seeds remaining in October				86%	55%	4%

seeds. These mature in May and June, but are found in largest numbers in October, mainly in Artemisietum.

When October soil samples were compared with those of June, there was a reduction in the numbers of seeds of six of the nine species which disperse their seeds at the beginning of summer. There was an increase in the numbers of seeds of the three synaptospermic species which disperse their seeds during summer.

3.2.6 Changes in the 'Seed Bank' of Plants That Disperse Their Seeds by Rain

Seeds that are dispersed by rain are released from the mother plants during the rains of the previous winter or winters. The seeds that did not germinate when separated from 120 g samples of crust taken from the three main habitats are indicated in Table 38. Samples from the soil crust, taken in June 1981, were compared with samples taken from the same habitats in October. Samples from Hammadetum contained 86%, while 55% were found in Artemisietum and only 4% in Zygophylletum samples (Table 38) (Gutterman et al. 1982).

Because seeds which are released and do not germinate are collected in very large numbers by seed-eating animals (Table 38), it appears that the protection afforded by dead mother plants is a very important factor in survival. The most efficient mechanism of seed dispersal by rain is found in *Blepharis* spp. and *Anastatica hierochuntica*. In these, (1) only a portion of the seeds is re-

Table 38a. Numbers of seeds of different plant species, whose seeds are dispersed by wind, separated and identified from 120 gm samples of soil from the soil crust layer collected in June and October 1981 from the three habitats: *Hammadetum* (H.s.), *Artemisietum* (A.h.a.) and *Zygophylletum* (Z.d.). Seeds from two samples from each habitat were identified. (After Gutterman et al. 1982)

Season	Number of seeds in samples collected in June 1981				Number of seeds in samples collected in October 1981			
Habitat	H.s.	A.h.a.	Z.d.	Total	H.s.	A.h.a.	Z.d.	Total
Plant species								
Filago desertorum	50	2	143	195	11	14	60	85
Roemeria hybrida[a]	2	4	2	8	2	68	–	70
Stipa capensis	–	3	15	18	–	4	–	4
Schismus arabicus	18	49	2	69	21	25	1	47
Cutandia memphitica	–	2	–	2	–	–	–	–
Trigonella arabica	2	–	1	3	–	2	–	2
Herniaria hirsuta[a]	–	2	–	2	2	28	–	30
Spergularia diandra[a]	–	–	1	1	1	11	–	12
Erodium hirtum	4	3	–	7	4	1	–	5
Total	76	65	164	305	41	153	61	255

[a] Synaptospermy species.

leased during a rain event and during one rainy season, (2) dispersal and germination may occur during the same rain event (Sects. 3.3.3.3, 3.3.3.4).

3.3 Seed Maturation in Summer and Dispersal by Rain in Winter

3.3.1 Plants Whose Seeds are Dispersed by Rain (Ombrohydrochory) and/or Runoff Water – The "Protection Strategy"

Species whose seeds are dispersed by rain (ombrohydrochory) have unique mechanisms of seed dispersal and germination. Seeds are protected from the time of maturation at the end of spring, and throughout the summer at least until the following winter rains in all the ombrohydrochoric species that occur in the deserts, as well as in arid habitats of the Mediterranean region of Israel. Seed dispersal by rain is especially important in extremely hot and dry deserts. In some of these not only is the percentage of species possessing mechanisms of synaptospermy (Murbeck 1919/20), "The feature in which two or more seeds or one-seeded fruits are joined to form a compound dispersal unit" (Zohary 1962), myxospermy "seeds or one-seeded fruits of many species when moistened, exude mucilage from their testa or pericarpial layers" (Zohary 1962), and ombrohydrochory higher than in other areas, but so is the number of individuals. The more extreme the desert conditions are, the greater is the number of plant species and the percentage of individual plants in the total plant population which disperse their seeds by rain.

Of 40 ombrohydrochoric plant species known in Israel and the Sinai Peninsula, 25 are derived from the Saharo-Arabian region or have entered from neighbouring regions (Tables 39, 40). These show up to seven different mechanisms of seed release during precipitation (see below). In most cases small, mucilaginous seeds adhere to the soil surface after dispersal and remain as a part of the soil crust until germination. Mass germination of seeds which accumulate in depressions and germinate after a long period of storage (such as *Mesembryanthemum nodiflorum, Aizoon hispanicum, Trigonella stellata*) may break the soil crust. The seeds of others germinate shortly after dispersal, especially under the most extreme desert conditions. Areas of several m^2, or even of several km^2 are, from time to time, almost completely covered with annuals such as *M. nodiflorum* (Gutterman 1980/81 a), *M. forsskalii, Aizoon canariense*, or *A. hispanicum* (Figs. 77, 78). Examples of this have been observed near the Dead Sea, one of the hottest and driest areas of Israel and where the soil is impoverished and saline. It is especially common in winters when rainfall is above average (Gutterman 1980/81 a).

3.3.2 Mechanisms of Seed Release by Rain

Ombrohydrochoric plants can be separated into seven different groups (Tables 39, 40) (Gutterman 1990 b).

Table 39. Some of the desert plant species divided according to their mechanisms of seed dispersal by rain (ombrohydrochory); their phytogeographic regions (area); mode of dispersal (Zohary 1966, 1972; Feinbrun-Dothan 1977, 1978; [1] see text). (After Gutterman 1990b)

Group number	Plant species	Family	Area	Mechanism of seed Dispersal by rain	
				Seed release	Mode of dispersal
1	Anthemis pseudocotula Boiss.	Asteraceae	Med., IT, SA	(C)	(A, S)
	Asteriscus graveolens (Forssk.) Less. [a]	Asteraceae	SA, SU	(C)	(A, S)
2	Anastatica hierochuntica L.	Brassicaceae	SA, SU	(B)(h)	(A, D, M)
	Carrichtera annua (L.) DC.	Brassicaceae	SA	(B)	(A, M)
	Reboudia pinnata (VIV.) O.E. Schulz	Brassicaceae	E.SA	(B)(R)	(A, M)(Z)(W)
	Erucaria boveana Coss.	Brassicaceae	E.SA (IT)	(B)	(A)
	Lepidium aucheri Boiss.	Brassicaceae	W.I.T.	(B)	(A, M)
	Alyssum damascenum Boiss. and Gaill.	Brassicaceae	W.I.T.	(B)(h*)	(A, S)
	Notoceras bicorne (Ait.) Caruel	Brassicaceae	SA	(B)(h*)	(A, S)
3	Aizoon hispanicum L.	Aizoaceae	SA	(H)	(A, S)
	A. canariense L.	Aizoaceae	SU	(H)(X)	(A, S)
	Mesembryanthemum forsskalii Hochst.	Aizoaceae	SU, SA	(H)	(A, S)
	M. nodiflorum L.	Aizoaceae	Med., W. Euro-Sib., SA	(H)	(A, S)
	M. crystallinum L.	Aizoaceae	Med., W. Euro-Sib., SA	(H)	(A, S)
3 a	Trigonella stellata Forssk.	Fabaceae	SA, W.I.T., Med.	(H)(X)	(A, P, S)
	Astragalus tribuloides Del.	Fabaceae	SA, I.T.	(H)(X)	(A, S)
	Astragalus asterius Steven	Fabaceae	SA, Med. (IT)	(H)(X)	(A, S)
	Neotorularia torulosa (Desf.) Hedge et Leonard	Brassicaceae	W.I.T. (SA)	(H)(X)	(A, S)
	Leptaleum filifolium (Willd.) DC.	Brassicaceae	IT, (SA)	(H)(X)	(A, S)
3 b	Filago contracta (Boiss.) Chrtek and Holub	Asteraceae	W.I.T., E. Med.	(H)*	(A, S)
	Anvillea garcinii DC.	Asteraceae	E. SA	(H*)(X)	(A, S)
	Gymnarrhena micrantha Desf.	Asteraceae	SA, I.T.	(H*)(G)	(D, W)(Z)
	Asteriscus pygmaeus (DC.) Coss. and Dur.	Asteraceae	SA, T.	(H*)(X)	(D, W)

Seed Maturation in Summer and Dispersal by Rain in Winter

4	*Plantago coronopus* L.	Plantaginaceae	Med., W.I.T., SA	(K)(h*)(X)(B)	(A,M)
5	*Blepharis attenuata*	Acanthaceae	W.I.T., E.SA	(K)(X)(B)	(D,M)
	B. ciliaris (L.) (B.L. Burtt)[a]	Acanthaceae	SA, SU	(K)(X)(B)	(D,M,F)
	B. linearifolia Pers.	Acanthaceae	SA	(K)(X)(B)	(D,M,F)
	Blepharis L. spp.[a]	Acanthaceae	SA	(K)(X)(B)	(D,M,F)

[a] Ecotypes of perennial species.

New Flora Names
Brassicaceae = Cruciferae (in Zohary 1966)
Fabaceae = Papilionaceae (in Zohary 1972)
Asteraceae = Compositae (Feinbrun-Dothan 1978)
Aizoaceae = Mesembryanthemaceae

Plant Geographical Regions
E.SA. = East Saharo Arabian; I.T. = Irano-Turanian; Med. = Mediterranean; S.A. = Saharo Arabian; SU = Sudanian; W. Euro-Sib. = West Euro-Siberian; W.I.T. = West Irano-Turanian.
(A) Atelechory; (B) fruits (capsules, siliqua or pods), with separation or dehiscence induced by rain; (C) capitula with crumbling induced by rain; (D) dispersal; (Z) the unopened terminal part of the pod (the beak of the pod) is dispersed by ants; (F) seeds with mucilaginous covers which are dispersed by floods in the wadis; (G) aerial dispersal units in an amphicarpous plant; (H) hydrochastic capsules or pods; (H*) hydrochastic capitula; (h) hydrochastic stems; (h*) hydrochastic peduncles, pedicels, rhachis; (K) hydrochastic sepals or bracts; (M) seed on which a mucilaginous layer appears on the seed coat after wetting; (P) pods that open apically after wetting and press the seeds out at the base of the pods; (R) one seed remains in the apical part of the pod; (S) very small or even larger seeds which stick to the wet soil surface immediately after dispersal by rain; (W) seeds with pappi which are dispersed by wind after having been released by rain or adhere by the pappi to the soil; (X) dead mother plants contain the seed bank of the species and release part of the seeds during years at events of rain.

Table 40. Ombrohydrochory of plant species of Israel, from the Mediterranean area and extending towards the Irano-Turanian region. (After Gutterman 1990b)

Group number	Plant species	Family	Area	Mechanism of dispersal
1	*Pallenis spinosa* (L.) Cass.	Asteraceae	Med., (I.T.)	(C)
	Phagnalon rupestre (L.) DC	Asteraceae	Med., W.I.T.	(C)
2	*Lepidium spinescens* DC	Brassicaceae	E. Med., (I.T.)	(h*)(B)
	L. spinosum Ard.	Brassicaceae	E. Med.	(h*)(B)
3b	*Cichorium pumilum* Jacq.	Asteraceae	Med., W.I.T.	(H*)(W)
	Asteriscus aquaticus (L.) Less.	Asteraceae	Med.	(H*)
4	*Plantago cretica* L.	Plantaginaceae	E. Med.	(K)(h*)(X)(B)(M)
	P. bellardii All.	Plantaginaceae	Med., W.I.T.	(K)(X)(B)
	P. crassifolia Forssk.	Plantaginaceae	Med.	(K)(h*)(X)(B)
6	*Salvia horminum* L.	Lamiaceae	Med., (W.I.T.)	(K)(h*)
	S. viridis	Lamiaceae	Med., (W.I.T.)	(K)(h*)
7	*Ammi visnaga* (L.) Lam.	Apiaceae	Med., W.I.T.	(U)[a]

[a] Hydrochastic umbel rays. For other abbreviations see Table 39.

Seed Maturation in Summer and Dispersal by Rain in Winter

Fig. 77. *Aizoon hispanicum* dry dead plant (height: 8 cm) in its natural habitat with the closed dry capsules containing mature seeds

Fig. 78. *Aizoon hispanicum* capsules. *On the left* dry closed capsule; *on the right* wet open capsule with seeds that will be dispersed by the following drops of rain (×3)

3.3.2.1 Seed Release by Capitula Crumbling Induced by Rain (C)

In *Anthemis pseudocotula, Asteriscus graveolens, Pallenis spinosa* and *Phagnalon rupestre* (Asteraceae), the seeds are protected from achene maturation at the beginning of the summer until the following winter. Drops of rain cause the capitula to crumble and release the achenes. These are then dispersed by wind (W) or adhere to the wet soil surface [atelechory (A)] until they germinate (Tables 39, 40) (Sect. 3.3.3.2).

3.3.2.2 Pod Dehiscence Induced by Rain (B)

Carrichtera annua (Figs. 57, 58), *Anastatica hierochuntica* (Fig. 79), *Reboudia pinnata, Erucaria rostrata, Lepidium aucheri, L. spinescens, L. spinosum, Alyssum damascenum* and *Notoceras bicorne* (Brassicaceae) have pods which separate (Tables 39, 40) along a dehiscence zone after wetting (Nordhagen 1936; Gutterman 1982a). These species have myxospermus seeds (Zohary 1937). The atelechoric (A) seeds are glued to the soil by their mucilaginous (M) layer (Tables 39, 40). In *Anastatica hierochuntica*, the hydrochastic woody branches (h*) (Fig. 79) close around the woody pods, protect them when they are dry and open when they are wet (Steinbrinck and Schinz 1908) (Sect. 3.3.3.3). In *R. pinnata* the beak, which is the unopened terminal part of the seed pod, is collected in large numbers by ants (*Messor ebeninus*) (Sect. 3.2.3) or dispersed by wind.

Fig. 79. *Anastatica hierochuntica. On the left* dry dead plant (size: 10 cm) in which the branches enclose and protect the pods on the inner side of the plant. *On the right* plant after wetting in which branches are opened and the capsules are weakened by a period of wetting which enables the following drops of rain to separate the valves and disperse the seeds. The seeds will adhere to the soil surface by the layer of mucilage surrounding them

3.3.2.3 Hydrochastic Capsules, Pods and Capitula

Aizoon hispanicum (Figs. 77, 78), *A. canariense, Mesembryanthemum forsskalii, M. nodiflorum* and *M. crystallinum* (Aizoaceae) possess hydrochastic capsules (H). These open after wetting and, on each heavy rain, at least some of the seeds within are dispersed by the rain drops (Fig. 80) (Gutterman 1980/81b, 1982b, 1990a,b; Ihlenfeldt 1983).

Trigonella stellata, Astragalus tribuloides, A. asterias (Fabaceae), *Neotorularia torulosa* and *Leptaleum filifolium* (Brassicaceae) are plants with hydrochastic pods (H). In *Trigonella stellata*, the apical part of the pod opens after wetting at the same time as the basal part of the pod pushes (P) out the seeds (Evenari and Gutterman 1976).

Filago contracta, Anvillea garcinii, Asteriscus pygmaeus, A. aquaticus, Gymnarrhena micrantha and *Cichorium pumilum* (Asteraceae) are equipped with an hydrochastic involucre of the capitula (H*) (Steinbrinck and Schinz 1908; Zohary and Fahn 1941; Fahn 1947, 1967; Fahn and Werker 1972).

Some of the species mentioned above are atelechoric (A), their small seeds adhering to the wet soil surface (S) and remaining as a part of the soil crust until they germinate. Others have achenes with pappi (Sect. 3.2.1.1) and are dispersed by wind [anemochory – (DW)]. Among these, there are species in which the dead mother plant is the main 'seed bank' which releases a portion of the seeds over a period of many years (Tables 39, 40) (Gutterman and Ginott, in press).

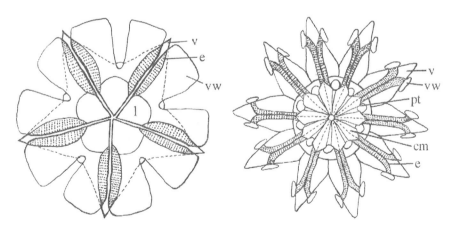

Fig. 80. Open capsule of the Aizoaceae: *on the left* a type (Delosperma) without covering membranes, seen from above; *v* valve, *vw* valve-wings; *l* locule; *e* expanding keel. *On the right* a type (Cephalophyllum) with covering membranes or cell lids (cm) and placental tubercles (pt). (Ihlenfeldt 1983)

3.3.2.4 Hygrochastic Bracts and Sepals, Separation of Capsules Induced by Rain

In *Plantago coronopus* (Figs. 72–74), the hydrochastic peduncles of the inflorescences bend towards the soil when they are dry, and straighten when they are wet (h*). The bracts and sepals (K) spread out after wetting; then the capsule breaks apart along the dehiscence zone (B). The mucilaginous seeds released (A, M) adhere to the soil surface (as in *Plantago cretica, P. bellardi* and *P. crassifolia*) (Tables 39, 40) (Zohary and Fahn 1941; Fahn 1947, 1967; Fahn and Werker 1972).

3.3.2.5 Hydrochastic Bracts, Sepals and Exploding Capsules

Blepharis spp. exhibits a double mechanism in having a combination of hydrochastic bracts and sepals (K) which cover the capsule when dry and open when wet. This exposes the capsule to wetting. After a period of wetting and of weakening of the 'lock area', the tension between the two halves of the capsules overcomes the strength of the lock, the fruit explodes along the dehiscence tissue (B) and the seeds are shot out (Gutterman et al. 1967) (Table 39; for more details see Sect. 3.3.3.4).

Two other modes of dispersal occur in the Mediterranean and Irano-Turanian areas (Zohary and Fahn 1941; Fahn 1947, 1967) (Table 40).

3.3.2.6 Hydrochastic Pedicels and Sepals

Salvia horminum and *S. viridis* (Lamiaceae) have a hydrochastic pedicel mericarp containing a calyx (h*). This opens after wetting and releases the seeds, which then adhere by their mucilage to the soil surface (Table 40) (Fahn and Werker 1972).

3.3.2.7 Hydrochastic Umbels and Rays

In *Ammi visnaga* (Apiaceae) the hydrochastic umbel rays (U) lock the fruits, after maturation, and release them by spreading after wetting (Table 40) (Gutterman 1990 b).

3.3.3 Differences in Dispersal Mechanisms of Some Groups of Species

Species of the same genus may have completely different mechanisms of seed release and different modes of dispersal. Examples from Sede Boker are *Trigonella stellata*, seeds of which are dispersed by rain, and *T. arabica* (Fig. 31) of which indehiscent pods are dispersed by wind and animals. Of the

Seed Maturation in Summer and Dispersal by Rain in Winter

40 species listed in Tables 39 and 40, ecological information is available only on the minority of species. More research needs to be undertaken for a better understanding of this unique adaptation for survival in the desert.

3.3.3.1 Differences in Dispersal Mechanisms of Some Species of the *Aizoaceae*

Aizoon hispanicum (Figs. 77, 78) and *A. canariense, Mesembryanthemum nodiflorum, M. forsskalii* and *M. crystallinum* are annual plants. In these five species, the hydrochastic capsules open after wetting, and the seeds within are dispersed by drops of rain. After desiccation, the capsules close again. *M. nodiflorum* has a complex set of mechanisms of seed dispersal and germination (Tables 17, 18). Together with other mechanisms, these enable this species to survive satisfactorily under extreme desert conditions in winters with above average precipitation (Gutterman 1980/81a, 1990a, b) (Sects. 2.2.7, 2.2.8).

Seed dispersal must (1) occur in suitable micro-sites (depressions); (2) only part of the seeds in one capsule may be dispersed during one fall of rain; (3) other groups of seeds may be released after subsequent periods of wetting. These groups show different degrees of germinability, according to their positions in the fruit (Table 17) (Gutterman 1980/81a).

Of ca. 2700 species of Aizoaceae in the Karoo, Little Karoo, Richtersveld, Namaqualand, South African and Namibian deserts, many have complicated additional structures. These include membranes that cover the seeds above, and 'placental tubercles', from the peripheral direction, in their hydrochastic capsules. By such devices the dispersal of the seeds is inhibited over a period of time, but the spatial dispersal is improved (Fig. 80) (Herre 1971; Court 1981; Ihlenfeldt 1983). In several species, *Pleiospilos bulosii*, N.E. BR., *Glottiphyllum linguiforme* N.E. BR., (Fig. 15) and *Cheiridopsis* N.E. BR., we have observed that seeds, still enclosed in their capsules, were able to germinate after the capsules had been kept moist for several weeks. Whether the dry capsule was still connected with the mother plant, or had been detached before, made no difference to germination. The membranes that cover the seed and prevent dispersal did not prevent the emergence of the cotyledons and the hypocotyls of the seedlings that had germinated in the capsule (Gutterman 1990a).

Two types of capsule, differing in size and seed number, have been found in *Glottiphyllum linguiforme.* There are central capsules beneath the shrub canopy, and peripheral capsules. The latter are easily disconnected. Seeds from both types show different degrees of germinability (Sect. 2.2.2; Fig. 15) (Gutterman 1990a).

In certain species of Aizoaceae (Mesembryanthemaceae), mericarps are formed containing a single seed in a pocket which is dispersed by the wind. The transition from a hydrochastic capsule with strong atelechoric features to an anemocharic dispersal of winged mericarps can be observed in this family through a series of closely related genera (Fig. 80) (Ihlenfeldt 1983; Gutterman 1990a, b).

3.3.3.2 Differences in Dispersal Mechanisms of Some Species of the Asteraceae

Among the species of the Asteraceae which inhabit the deserts of Israel and whose seeds are dispersed by rain during the following winter the capitula of *Anthemis pseudocotula* and *Asteriscus graveolens* are induced by rain to crumble. The achenes are dispersed during the wet season after being protected in closed capitula throughout the summer. In *Filago contracta* and *Anvillea garcinii* involucre bracts of the capitula are hydrochastic. They open after wetting and close when dry. Almost all the achenes are dispersed during the first season of rain. In these species, the achenes are protected against seed-eating animals, not only during the first summer but also during the following rainy season, between one fall of rain and another. *Gymnarrhena micrantha* is an amphicarpic annual in which the involucral bracts of the aerial capitula (Fig. 81) open after wetting and the achenes, with their large pappus, are dispersed by wind (Fig. 14). In this species the subterranean achenes remain in the soil protected by the woody remains of the dead mother plant (Engler 1895; Warburg and Eig 1926). They germinate in situ, in the micro-habitat that was satisfactory for the mother plant (Zohary 1937; Koller and Roth 1964). In this species, all achenes of the aerial inflorescences are released when rain falls, and accumulate above the dead mother plant with their pappi open. When not

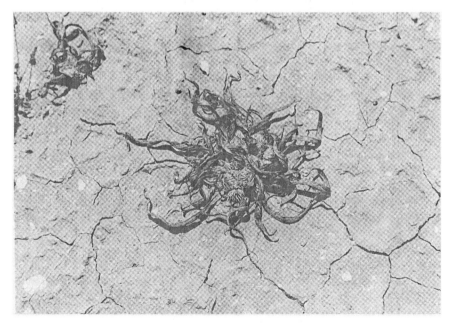

Fig. 81. *Gymnarrhena micrantha* plant in its natural habitat on a flat loess plain, in summer. The plant is dead, leaves are dry and the aerial capitula remained closed until the next rainy season. (Size: 4 cm)

carried away by the wind after release, a large percentage are harvested, particularly by *Messor ebeninus* and *M. arenarius*. It is interesting to note that the activity of these ants increases a few hours after rain (Hord 1986) so that they can collect the seeds of ombrohydrochoric plants which are still free on the soil surface. In the Chihuahuan desert of North America, harvester ants also increase their activity shortly after rain (Whitford 1978).

Seed dispersal and germination of *Asteriscus pygmaeus* are regulated by rainfall long after the plant has dried out. The achenes of this annual are protected by woody hydrochastic capitula bracts. Each season, some of the achenes disperse when rain falls but the main seed bank is protected in the capitulum for many years. *A. pygmaeus* inhabits the hottest and driest habitats around Sede Boker including the south- and east-facing hill slopes (Fig. 82) and is fairly common in the other deserts of Israel including the Judean desert, north and central Negev, the lower Jordan Valley, the Dead Sea and Arava valley. The distribution of this species is Saharo-Arabian, extending somewhat into the Western Irano-Turanian geographic region (Feinbrun-Dothan 1978). Germination takes place during winter. This species shows a photoperiodic facultative long-day response for flowering. Regulation of the

Fig. 82. *Asteriscus pygmaeus* plant in flower during spring surrounded by capitula of dead mother plants which are the main seed bank of this plant for many years. Each season of rain some achenes are released, germinate and develop plants in almost the same micro-habitat. (Gutterman and Ginott, in press) (×1)

Table 41. *Asteriscus pygmaeus* inflorescence diameter; number of achenes arranged in number of whorls. (After Gutterman and Ginott, in press)

Whorl number	No. of achenes per whorl		
	6 mm	9 mm	12 mm
1	20	22	30
2	14	20	36
3	8	18	26
4	6	18	24
5	4	16	18
6		12	18
7		10	16
8		10	12
9		10	10
10			10
11			10
Total no. of achenes per inflorescence	52	136	210

flowering time and life cycle are dependent on day length, provided there is enough water in the soil. This ensures that the plant will complete its life cycle before the hot, dry summer. The later germination occurs in the season, the longer the day length under which development takes place. This results in a shorter time before the appearance of flower buds. There is a gradual reduction in the number and size of leaves and, as the days lengthen, the age at which the first buds appear (Evenari and Gutterman 1966; Gutterman 1982a). Each stem terminates in an inflorescence. Under short day lengths the capitula on the main stems may have a diameter of 16 mm and contain about 210 achenes, arranged in 11 whorls. The size of the capitula is much smaller when flowering takes place in long days than when it does in short days. Some of the inflorescences on the lateral stems are very small (6 mm) with only 50 achenes arranged in five whorls (Table 41) (Gutterman and Ginott, in press). The inflorescence (capitulum, head) is covered with a woody structure of about 11–18 upper leaves (bracts) arranged in two whorls. At the end of spring, during maturation of the achenes, the upper parts of the bracts dry up and are shed.

The achenes remain enclosed by the woody bracts for many years, and form the main 'seed-bank' of this species. Each season, some of the achenes are dispersed by rain. During the first summer after maturation there are changes in the capitulum that enable some of the achenes to be dispersed during the following winter. Over the years, the shape of the inflorescence changes according to the number of achenes that have been dispersed from the peripheral zone. Finally, a stage is reached when only a small group of achenes remains in the centre of the receptacle (Fig. 83). By this time, the colour of the inflorescence is light-grey in comparison with yellowish-brownish during the first

Fig. 83. *Asteriscus pygmaeus* opened capitulum *on the right* with a group of achenes that have not yet been dispersed, *at the centre*. *On the left* a dry closed capitulum (size: 1 cm). (Gutterman and Ginott, in press)

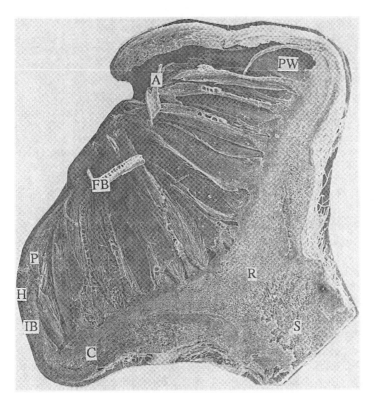

Fig. 84. Scanning electron microscope (S.E.M.) photograph of median cross-section of an *Asteriscus pygmaeus* inflorescence (size: 2 cm), showing the plant stem (*S*), receptacle (*R*), cohesion tissue (*C*), inflorescence bract (*IB*), flower (achene) bract (*FB*), achene (*A*), peripheral whorl of achenes (*PW*), central whorl of achenes (*CW*), achene pappus (*P*), hydrochastic cohesion tissue (*H*)×13. (Gutterman and Ginott, in press)

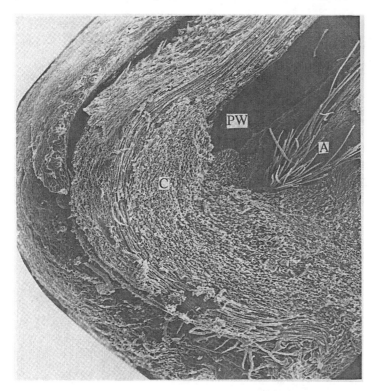

Fig. 85. S.E.M. photograph of part of median cross-section of an *Asteriscus pygmaeus* inflorescence showing achene (*A*), inflorescence bracts of the two whorls peripheral achenes whorl (*PW*), cohesion tissue at the base of the bract (*C*). ×75. (Gutterman and Ginott, in press)

summer, which turns to brown later on. The older the inflorescence, the darker the colour. Throughout these stages, the bracts open only a few minutes after wetting. In very old inflorescences, however, from which the achenes have already dispersed, the mechanism no longer works and the bracts are shed.

During the first summer after their formation the bracts may open more than 90 times when dew is deposited and even when the humidity is more than 85%. Cohesion tissue, situated in the receptacle, swells when wet and shrinks when dry. This causes some achenes of the peripheral whorl to become disconnected from the receptacle (Figs. 84–86). This action also engenders a knee-like bend at the middle of the bracts where the hydrochastic tissue exists (Figs. 87–90). Another bend at the base of the bract where cohesion tissue also exists, encloses the achenes very tightly in the inflorescence and protects them from seed predators. Constant opening and closing affects the degree to which the bracts open in response to rain from 90° to about 130°. This enables further drops of rain to cause the achenes already disconnected, to drop out (Gutterman and Ginott, in press).

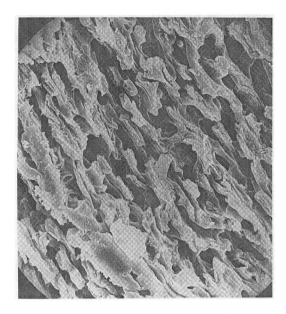

Fig. 86. S.E.M. higher magnification (×700) of part of the cohesion tissue of the inflorescence receptacle of *Asteriscus pygmaeus*. (Gutterman and Ginott, in press)

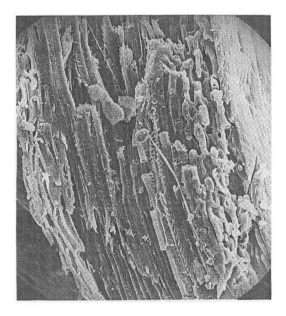

Fig. 87. S.E.M. photograph of median cross-section of an *Asteriscus pygmaeus* inflorescence bract, at the knee-like bend area, showing adaxial hydrochastic cohesion tissue *on the right* and counterforce abaxial fiber tissue *on the left.* ×140. (Gutterman and Ginott, in press)

Dew no longer causes the bracts to open after the inflorescences have been exposed to rain or have imbibed water over long periods. From the next winter, only rain can induce opening (Fahn 1947). Some of the achenes which have been disconnected, are then dispersed by rain (Koller and Negbi 1966). The pappus of the achenes opens after wetting when the achenes and their bracts

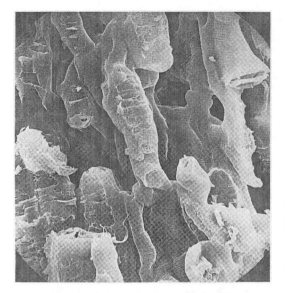

Fig. 88. S.E.M. high magnification (×616) of the hydrochastic cohesion tissue of *Asteriscus pygmaeus* (Fig. 87) showing the transversal structure arrangement of the cell walls which enable them to swell in the longitudinal direction. (Gutterman and Ginott, in press)

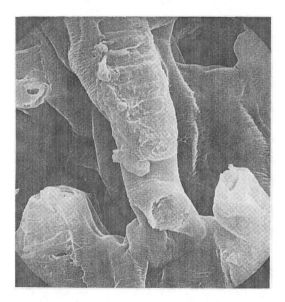

Fig. 89. S.E.M. higher magnification (×1232) of some cells of the hydrochastic cohesion tissue (Fig. 88) of *Asteriscus pygmaeus*. (Gutterman and Ginott, in press)

are disconnected from the receptacle but still situated in the inflorescence. After dispersal, the short pappus of the achenes can adhere to the wet soil surface so that the achenes remain near the dead mother plant and germinate there. Other achenes are carried away by wind or runoff water and germinate in runnels. Some are collected by ants and other seed eaters. Whorl after whorl of achenes are dispersed: the peripheral ones, two to five per season of rain, are the first to be dispersed. The central achenes are the last to go.

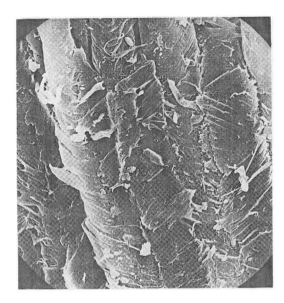

Fig. 90. High magnification (×1120) of counterforce tissue of *Asteriscus pygmaeus* (Fig. 87) showing fibre cell walls with micro-lamellae of the cell wall in layers, lying in different directions. (Gutterman and Ginott, in press)

Disconnected peripheral achenes have been found to germinate to a significantly higher level than achenes from the peripheral and sub-peripheral whorls disconnected manually just before imbibition. Germination of the sub-peripheral achenes was significantly lower, even after 60 days of imbibition. Achenes from the whorls near the centre did not germinate at all (Fig. 91) (Gutterman and Ginott, in press). It is possible that achenes may contain a water-soluble inhibitor: immersion in water for 20 h as pretreatment increases germination from 4 to 52%. The achenes need to be disconnected from the capitulum before they germinate (Koller and Negbi 1966). This effect of position ensures that achenes of each whorl germinate when they reach the periphery and are disconnected. The advantage is that, instead of being dispersed by wind at the beginning of summer, immediately after maturation, the seeds are protected until the following winter or winters. Only a few achenes are then released and dispersed by rain and possibly by wind over a period of many years.

3.3.3.3 Differences in Dispersal Mechanisms of Some Species of the Brassicaceae

In *Anastatica hierochuntica* (Fig. 79) both local dispersal on the spot by rain, and distant dispersal by running water have been demonstrated (Friedman and Stein 1980). About 1.5 h after wetting, the dehiscent zone between the two valves of the pod becomes weak enough for an additional drop of rain to separate the pod with its two upright teaspoon-like extensions. This causes dispersal of the mucilaginous seeds. Germination of *A. hierochuntica* seeds may take place within 6 h in favourable conditions.

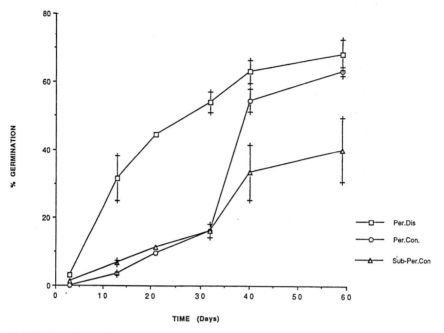

Fig. 91. *Asteriscus pygmaeus* achene germination (%±S.E.) during 60 days of imbibition of achenes previously disconnected (*Per.Dis.*) from peripheral whorl; manually disconnected from peripheral whorl (*Per.Con.*) and from the sub-peripheral whorl (*Sub-Per.Con*). (Gutterman and Ginott, in press)

Carrichtera annua (Figs. 57, 58) is another ombrohydrochoric plant (Nordhagen 1936). During winter, after about 1.5 h of continuous rain or continuous wetting of the pod (siliqua), the dehiscence zone between the two halves of the pod becomes weakened. When another drop of rain falls on the teaspoon-like extension of the siliqua, the pod opens, the mucilaginous seeds fall and adhere to the soil near the dead mother plant (Figs. 131, 132). In this case, at least part of the seeds will germinate within a short time after dispersal (Gutterman 1981). Such topochoric seed dispersal (Burmil 1972) ensures that *C. annua* is distributed in patches. Distant establishment of new patches can occur when an entire dry plant is broken off during the first summer and carried a long distance by the wind (Gutterman 1982b). The wet mucilaginous seeds are able to adhere to the bodies of mammals or to the beaks and legs of birds, and are thus carried for even longer distances.

3.3.3.4 Differences in Dispersal Mechanisms of Some Species of the Acanthaceae

The genus *Blepharis* (Acanthaceae) includes annual and perennial species as well as many ecotypes (Figs. 92, 93) (Gutterman et al. 1967, 1969a; Gutterman

Fig. 92. *Blepharis linearifolia*, the annual desert plant, from a population on Tiran Island, grown under greenhouse conditions. This shows the dichotomic branching appearing from the leaf axil of two out of four leaves situated beneath the inflorescence, which is typical of the species of the genus *Blepharis* of Israel and the Sinai Peninsula

Fig. 93. A shrub of *Blepharis ciliaris* situated on a wadi bank in its natural habitat in Nahal Tamar, 10 km south of Sodom

1972, 1989h). All of these have a double safety mechanism for seed dispersal. This ensures that only some of the seeds will be dispersed in one precipitation and then only after a minimal amount of continuous wetting has taken place. This double safety mechanism acts in two stages; the first is the opening by wetting of the sepals which enclose the capsules (Fig. 94); opening to a distance

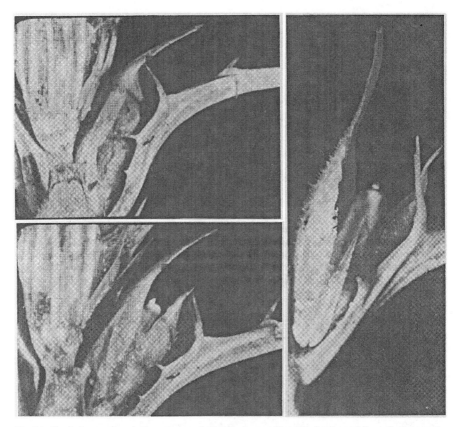

Fig. 94. Single bract of an inflorescence of *Blepharis* spp. *Top left:* in the axil is a single capsule enclosed by the dry persistant sepals. ×5; *Bottom left:* Separation of the upper and lower sepals after short moistening; *Right:* Further separation of sepals of *Blepharis* spp., in response to wetting exposes the capsule to rain. ×6.3. (After Gutterman et al. 1967)

Table 42. Separation of sepals induced by wetting of *Blepharis* spp. Range in mm of distance between upper and lower sepals. (After Gutterman et al. 1967)

Time	5 min	10 min	15 min	22 min	47 min	160 min
Green branches	0	0−3[a]	0−3	2−3	3	5
Brown branches	1−3	3−5	3−5	6	−	−

[a] Range in mm.

of 5 mm takes 10−160 min continuous wetting when submerged in water. The older the inflorescence, the quicker will it open (Table 42) (Gutterman et al. 1967). The second stage begins when rain continues to fall. The 5 mm gap enables drops of rain to wet the 'lock area' which is located on the top of the capsule (Fig. 95). Ten minutes continuous wetting is necessary for 32% of

Seed Maturation in Summer and Dispersal by Rain in Winter 125

Table 43. Time course of water imbibed capsule dehiscence of *Blepharis attenuata* as a function of capsule age. (After Gutterman et al. 1967)

Origin of capsules	% Capsules dehisced after indicated time of imbibition in minutes										
	5	10	15	20	30	40	50	60	90	120	180
Capsules from green branches	0	2	6	8	26	36	48	52	52	52	52
Capsules from dry brown branches	0	32	32	36	72	80	80	80	80	84	92
Capsules from green branches, apex scarified	86	88	–	–	–	–	–	–	–	–	100
Capsules from dry brown branches, apex scarified	100	–	–	–	–	–	–	–	–	–	–

capsules from dry brown branches, and about 40 min for 36% of capsules from green branches, to explode and disperse their seeds. Explosion occurs when water enters and weakens the 'lock area'. Tension between the two halves of the capsule then overcomes its strength and that of the dehiscence zone (Figs. 95 – 99; Table 43). In species that inhabit wadi beds, the exploding capsule causes the seeds to be shot out to a distance of several metres if dispersal is caused by rain, and sometimes to distances of more than 100 m when the dispersal occurs as a result of flooding (Gutterman et al. 1967). This double safety mechanism ensures that seeds are dispersed only after a long period of wetting or by flood. Furthermore, even under optimum conditions (Table 44), only about 30% of capsules actually disperse their seeds. This is because water may soak into the peripheral fibre bundles of the capsule, thereby releasing the tension of the capsule (Table 44). In capsules with scarified apices, water penetrates the 'lock area' more easily than other parts of the capsule, which are sealed with paraffin. The result is that 90% of the capsules dehisce within 5 min. A single hole in the side of a capsule enables water to penetrate the peripheral fibre bundles and dehiscence is prevented. In nature, tension develops again during the following dry period. This ensures that seed dispersal takes place only after a certain amount of rain has fallen and that only a portion of the seeds will be dispersed at one time. This mechanism renews itself between one rain event and another (Gutterman et al. 1967; Gutterman 1990 b).

In *Blepharis* spp., the 'seed bank' remains on the dead mother plant. The dry seeds (Figs. 99, 100) with multicellular hairs that lie along the seed surface and cover the coat (Fig. 101). When they fall onto wet soil the hairs open and buliform-like cells swell when wetted (Figs. 102, 104). The seed then changes its angle so that it lies at 30 – 45 ° to the soil surface and the micropyle touches the soil (Figs. 103, 108). The cells at the top of the hairs (Figs. 104, 105) separate and adhere to the soil particles (Figs. 106, 107). The seed thus becomes connected by the hairs to the soil. As a result of this, the root is able to penetrate the soil. The seed germinates in a wide range of temperatures from 8 – 40 °C in both light and dark (Table 45). In some conditions, the root may

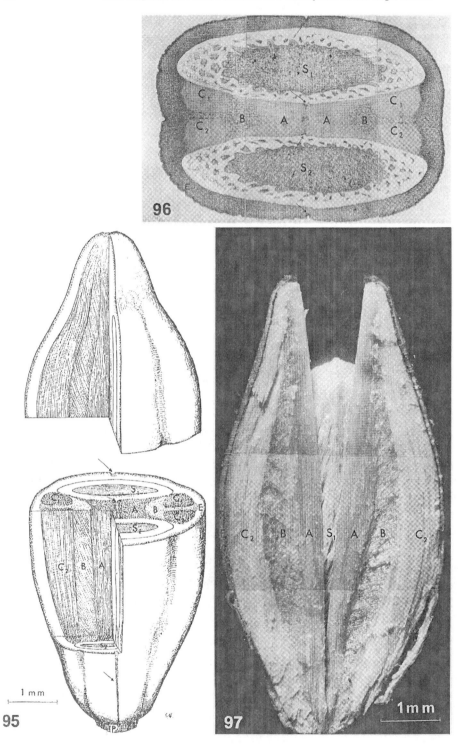

Seed Maturation in Summer and Dispersal by Rain in Winter

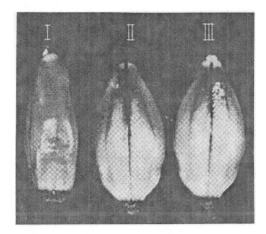

Fig. 98. Entire capsules of *Blepharis* spp. *I* Side view; *II* lower side; *III* upper side. ×5. (After Gutterman et al. 1967)

Fig. 99. Capsule of *Blepharis* spp. opened along the lines of dehiscence showing two seeds and their respective hook-like jaculators. ×5. (After Gutterman et al. 1967)

◄───

Fig. 95. Schematic drawing of a *Blepharis* spp. capsule indicating the longitudinal (Fig. 97) and cross-sectional (Fig. 96) anatomy. Each half of the septum is composed of two fibre bundles (*A* and *C*) separated by a parenchyma bundle (*B*). A narrow projection of parenchyma bundle *B* separates fibre bundle *C* into two parts, C_1 and C_2. The septum splits between the two inner fibre bundles (*A* and *A*). Seeds S_1 and S_2 are separated by the septum. *Arrows* indicate dehiscence zones in capsules wall (*E*) and in the septum. *P* indicates the pedicel. (Gutterman et al. 1967)

Fig. 96. Cross-section of nearly ripe capsule of *Blepharis* spp. The two seeds (*S*) are separated by the septum composed of fibre bundles *A*, C_1 and C_2. Between the fibre bundles is parenchyma bundle *B*. *Arrows* indicate dehiscence zones in capsule wall and in the septum. ×35. (After Gutterman et al. 1967)

Fig. 97. Longitudinal-section through the various tissues of the septum in a mature capsule of *Blepharis* spp. *A* and *C* are fibre bundles, *B* is the parenchyma, *S* is the dry seed covered by multicellular integumentary hairs. The septum has split between the two fibre bundles *A*. The capsule wall (*E*) has shrunk during maturation. ×22.3. (After Gutterman et al. 1967)

Table 44. Dehiscence between *Blepharis ciliaris* capsules on imbibition after treatments by scarification, a hole at the side of the capsule and/or a cover of paraffin. (After Gutterman et al. 1967)

Treatment			% Capsules dehisced after:			
			5 min	15 min	95 min	25 h
1		Whole capsules entirely coated with paraffin	0	0	0	0
2		Whole capsules, covered with paraffin only at the apices	0	0	0	0
3		Capsules with scarified apices but apices covered with paraffin	0	0	0	0
4		Capsules with scarified apices and all but the apex coated with paraffin	90	100	–	–
5		Capsules with scarified apices without paraffin	70	90	100	–
6		Whole capsules covered with paraffin except for base of capsule	0	0	0	0
7		Whole capsules no paraffin (control)	0	0	30	30
8		Capsules with a single hole in side wall	0	0	0	0
9		Capsules, same as 8 but entire capsule coated with paraffin except for hole	0	0	0	0
10		Same as 8 except that apices were coated with paraffin	0	0	0	0

reach a length of 49 mm within the first 24 h after wetting, and the growing cotyledons protrude from the integument. The integument remains connected to the soil by its multicellular hairs after the rapid germination has taken place (Fig. 109) (Gutterman 1972), unless there is excess water present (Gutterman et al. 1967; Witztum et al. 1969).

When seeds of *Blepharis ciliaris*, which inhabit wadi beds, imbibe water, a layer of mucilage a few millimetres thick, is developed within a few minutes (Figs. 110, 111). This prevents germination and protects the embryo from damage by flood. Such is not the case with seeds of *B. attenuata*, whose seed

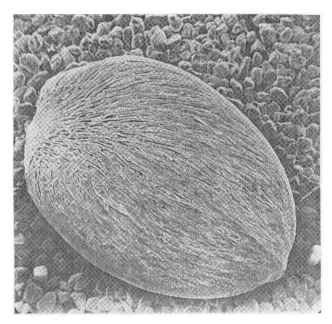

Fig. 100. Dry seed of *Blepharis* spp. with adpressed multicellular hairs. ×12. (After Gutterman et al. 1967)

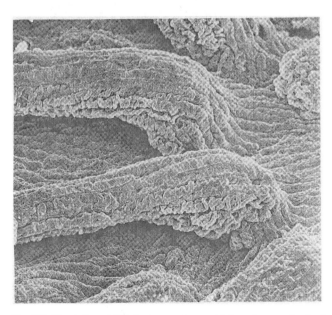

Fig. 101. Dry hairs of *Blepharis* spp. dry seed lying along and adhering to the seed integument (×330)

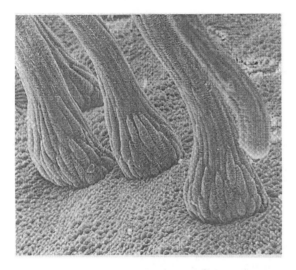

Fig. 102. The opening of the hairs to the angle of 90° after wetting. ×75. (Gutterman and Witztum 1977)

Fig. 103. Side view of the stages in the raising by hairs to an angle of 40–45° of a seed as a function of the hydration of the multicellular integumentary hairs. The mucilaginous hairs bind the seed and future seedling to the soil surface. This sequence took less than one minute. ×5. (Gutterman et al. 1967)

coats contain much less mucilage: this species inhabits hill slopes in the Judean desert and cannot be found in wadis (Tables 46, 47). The mucilaginous layer is composed of cellulose microfibrils (Fig. 112) in an uronic gel (Figs. 110, 111) and develops from the cell walls of the multicellular hairs (Figs. 104–107) and integument. This layer also prevents oxygen from penetrating to the embryo (Table 48) by convective water flow. The O_2 that reaches the embryo by diffusion is insufficient for germination, which is thereby inhibited until the excess of water disappears. The delay in germination prevents damage to the embryo and the radicle from turbulent flooding, and enables the seed to be carried unharmed by flood water for long distances (Gutterman et al. 1967, 1969a; Witztum et al. 1969).

Species and ecotypes of *Blepharis* that are not native to wadi beds, do not develop heavy layers of mucilage around the seed coat, and their seeds ger-

Fig. 104. Seed wetted and stained with Nile Blue, photographed under water to show the open multicellular integumentary hairs. ×8. (After Gutterman et al. 1967)

Fig. 105. Tip of a single multicellular integumentary hair showing portions of many cells with helical secondary walls which uncoil upon wetting. This is a magnification of the tip of a single hair shown in Fig. 104. ×128. (After Gutterman et al. 1967)

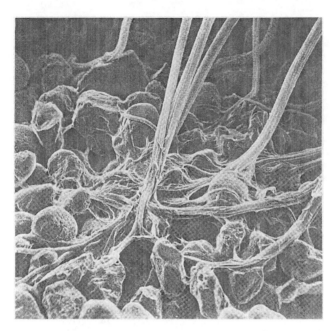

Fig. 106. At the tip of multicellular *Blepharis* hairs mucilaginous cells separate and glue together the soil particles to which the hairs become attached, anchoring the germinating seed to the soil. ×85. (After Gutterman and Witztum 1977)

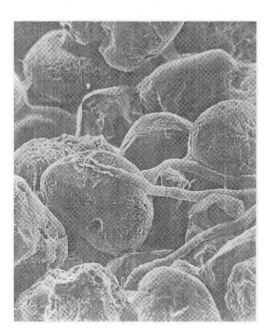

Fig. 107. Magnification of separated cells of hairs attached to the soil particles. ×650. (After Gutterman and Witztum 1977)

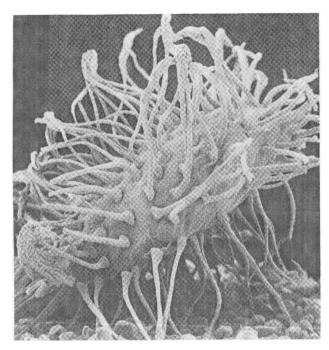

Fig. 108. The whole seed with the open multicellular hairs at an angle of 45°. The root tip is attached to the soil surface and anchored to the soil by the mucilaginous ends of the hairs. ×12. (After Gutterman and Witztum 1977)

minate well even when they are below water and in constant agitation (Gutterman et al. 1967, 1969a; Witztum et al. 1969).

It is interesting to compare the species of the desert genus *Blepharis* with the *Acanthus syriacus*, also a member of the Acanthaceae, which inhabits the East Mediterranean zone and West Irano-Turanian area of Israel (Feinbrun-Dothan 1978). In the latter species the capsules have the same anatomical arrangements, but the tension which develops when the capsule dries up during maturation overcomes the strength of the tissue in the 'lock area', and seeds, with hard seed coats, are shot out by the exploding capsule. This occurs during the summer, immediately after maturation. The sepals of *Acanthus* are not hydrochastic, as are those of *Blepharis* spp. The hydrochastic sepals and the small difference in relative strength of the 'lock area' in the capsules of *Blepharis* bring about dramatic differences in the seed dispersal strategy of the two genera (Gutterman 1990b). The small difference that develops in *Blepharis* spp. ensures seed dispersal and germination at the right moment in areas with unpredictable falls of rain.

There are about 100 species of *Blepharis* (Engler 1964) in addition to many ecotypes that are well adapted to desert conditions even in deserts that receive rain during summer. These include the southern part of the Sahara, and the

Table 45. Germination % and roots length (mm) after 24 h in constant temperatures from 8–45 °C of seeds of *Blepharis* spp. from Tiran Island and Mishor Rotem, at light and dark conditions (±S.E. of 25 seeds and seedlings). (Gutterman 1972)

Germination temperatures (°C)	Seeds from Tiran Island				Seeds from Mishor Rotem		
	Germination % after 24 h in light	Root length (mm) after 24 h in light	Germination % after 24 h in dark	Root length (mm) after 24 h in dark	Germination % after 24 h in dark	Root length (mm) after 24 h in dark	
8	–	–	–	–	60	1.8±0.7	
10	100	6.2±0.2	100	4.6±0.3	100	9.6±0.7	
15	–	–	–	–	100	9.6±2.5	
20	100	27.1±0.6	100	–	100	17.6±1.6	
25	100	31.2±1.5	100	32.5±2.2	100	20.4±2.8	
30	100	42.0±0.9	100	48.7±2.4	80	5.0±1.3	
35	100	34.8±3.0	100	42.4±3.5	40	–	
40	100	8.6±1.0	63.2	2.8±0.7	20	–	
45	0	0	0	0	0	0	

Table 46. Response of whole seeds and embryos of *Blepharis attenuata* and *B. ciliaris* to different water levels after 22 h. (After Gutterman et al. 1967)

Plant species and ecotypes	Seeds or embryos	Amount of water or depth of water	Seeds in which the roots elongated (%)	Seeds with geotropic root response [b] (%)	Average root length (mm)
Blepharis attenuata	Seeds	1 ml	100	100	22.5
	Seeds	3 ml	70	85	1.9
	Seeds	3 cm	10	55	0.3
	Embryos [a]	3 cm	100	100	6.2
Blepharis ciliaris	Seeds	1 ml	70	70	12.6
	Seeds	3 ml	0	60	0
	Seeds	3 cm	0	30	0
	Embryos [a]	3 cm	100	100	3.8

[a] Seed coat removed.

[b] But no emergence of roots from the mucilage layer.

Seed Maturation in Summer and Dispersal by Rain in Winter

Fig. 109. A seedling 24 h after 24 h of wetting. (After Gutterman et al. 1967)

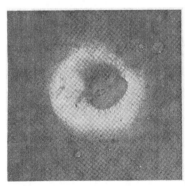

Fig. 110. Mucilage from seed of *Blepharis* spp. placed under distilled water for 24 h forming colourless halo around seed when placed in aqueous solution of Bismarck Brown. ×2. (After Gutterman et al. 1967)

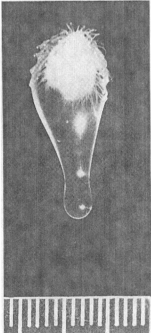

Fig. 111

Fig. 111. *Blepharis* spp. seed after imbibition of distilled water: five seeds per 30 ml, at 4 °C for 22 h. (After Witztum et al. 1969)

Table 47. Percentage germination of seeds of *Blepharis ciliaris* which inhabits wadis, and of *B. attenuata* which only inhabits slopes, in 7 cm jars under 3 and 9 cm water, 25 seeds per jar, after 12 h. (After Gutterman et al. 1967)

Species	Germination (%)		
	Volume 120 ml 3 cm depth	Volume 400 ml 9 cm depth	Volume 400 ml 9 cm depth with constant shaking
B. attenuata	0	0	100
B. ciliaris	0	0	0

Saudi-Arabian desert as well as some of the desert areas of South Africa. Other ecotypes inhabit deserts where rain falls during winter, such as the northern part of the Sahara, the Sinai Peninisula, and the deserts of Israel, Jordan, north Saudi-Arabia, Iraq and Iran. *Blepharis* plants cover very wide areas of the Sahel, in the southern part of the Sahara desert, among which there are very important plants for grazing. In years that *Blepharis* plants do not appear, the sheep and camels cannot find enough food to eat. In this area it has been observed that dead parts of these plants stick to the coats of passing mammals, such as goats and camels, by means of the spiny bracts on the inflorescences, and are thus dispersed (exozoochory) (Zonneveld, pers. comm.). Cloudsley-Thompson (1968) found near Omdurman and Khartoum (at the junction of the Blue and White Nile) that *Blepharis ciliaris* (*edulis*) was restricted to the slopes of the hills because they were 'too selectively grazed elsewhere by camels'.

Blepharis species from Sinai, Tiran Island, Israel and S. W. Iran have been found to be day-neutral for flowering (Gutterman 1988b, 1989h). This enables them to develop flowers and produce seeds every time there is sufficient rainfall, regardless of season. The higher the temperature and water stress, the

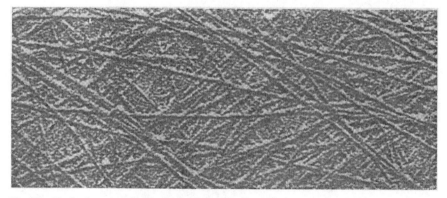

Fig. 112. *Blepharis* spp. seeds. Mucilage shadowed with platinum. ×69,000. (After Witztum et al. 1969)

Table 48. Germination, root growth and cotyledon colour of *Blepharis ciliaris* in 1 or 3 ml water in an atmosphere of air or oxygen. (After Gutterman et al. 1967)

Atmosphere	Amount of water (ml)	Germination (%)	Average root length (mm)	Cotyledon colour[a]
Air	1	100	11.9	Greenish Yellow
	3	0	0	Yellow
Oxygen	1	100	17.3	Green
	3	96	13.3	Greenish white

[a] The greening of the cotyledons is used as a bioassay of the relative amounts of O_2. No O_2, no greening; more O_2, more greening.

shorter the time until the appearance of flowers, from a minimum of $14-22$ days under high temperatures and water stress to $81-121$ days under more favourable conditions for growth and development. Rainfall, temperature and the soil humidity are the main factors which influence plant development and renewal in desert regions (Gutterman 1988b, 1989h).

To summarize: the double safety mechanisms of populations of *Blepharis* spp. that have been observed in the deserts of Israel, Tiran Island, the Sinai Peninsula, E. Egypt, S. W. Iran and South Africa, ensure that: (1) the relatively large seeds ($5.2-6.7$ mm $\times 3.6-4.4$ mm $\times 1$ mm) are protected from seed eaters by the woody sepals and capsules, from the time of maturation until the fall of rain during which the seeds disperse and germinate. This protection is especially important under extreme desert conditions where food for animals is limited and seeds are the main source of concentrated food; (2) only a portion of the seeds are released from the mother plant and germinate at one time, thereby spreading the risk; (3) seeds germinate at the right time after a fall of rain. This gives the seedlings a good chance to establish themselves; (4) seeds germinate in the range of $8-40\,°C$, in both light and dark. Within 24 h the root may grow to a length of about 49 mm; (5) an excess of water inhibits the germination. This enables seeds to be dispersed by flood without damage to the embryo or the radicle; (6) seeds germinate in suitable sites or ecological niches, such as wadi banks, after excess water has disappeared. When they are dispersed by flood they may be carried for a long distance from the mother plant. When they are dispersed by rain they do not travel so far (Gutterman et al. 1967, 1969a; Witztum et al. 1969; Gutterman 1990b); (7) *Blepharis* spp. and its ecotypes inhabiting the Negev and Sinai are day-neutral for flowering. Consequently, they can flower and produce seeds every time there is sufficient rainfall, regardless of the season. Flowers appear first under conditions of high temperatures and water stress after only $14-22$ days. In better conditions the first flowers appear after $81-121$ days; (8) rainfall, temperature and soil humidity are the main factors affecting seed dispersal, germination and flowering in *Blepharis* spp.; (9) these plants, with spiny bracts on their inflorescences may become attached to the coats of mammals and be dispersed in that way (exozoochory). The mucilaginous seeds may also become stuck to legs and beaks of birds and be dispersed to greater distances.

Mechanisms common to plants whose seeds are dispersed by rain are important in protecting the seeds from the time of maturation at the end of the growing season and throughout the summer, when seed-harvesting insects are active. Seed dispersal triggered by rainfall is beneficial because the activity of seed predators is minimal during times of rain.

3.4 Seed Maturation in Summer and Germination in situ (Atelechory) in Winter

Seeds that germinate in situ and do not disperse are found in some amphicarpic plant species. One example is *Gymnarrhena micrantha* (Fig. 81), which possesses subterranean achenes that remain protected under the soil surface until they germinate. The aerial dispersal units are released by rain and afterwards are dispersed by wind (Zohary 1937). Some of them are also dispersed by ants. The subterranean propagules of the amphicarpic plant *Emex spinosa* (Evenari·et al. 1977) also germinate from the dead mother plant. The subterranean achenes or propagules are well protected against predation, but at the cost of not being dispersed. This results in competition between plants in favourable micro-habitats (Sect. 2.2.1).

3.5 Seeds That Mature and Are Dispersed in Winter

Artemisia sieberi (Asteraceae) is one of the dominant plants on the north-facing slopes of the Central Negev and on all the slopes of the Irano-Turanian region at elevations of 800 – 1000 m. The mucilaginous achenes mature and are dispersed during December and January, the months when rain is most likely to fall. After dispersal, the seeds adhere to the soil surface (Fig. 139) until they germinate after a minimum of 16 days of continuous wetting. They need light for germination, and covered achenes will not germinate. These unique conditions for germination occurred in the Negev desert in 1963/64. They engendered a mass germination of 31.3 seedlings per m^2. These appeared in large areas of the Negev Desert highlands. In other years, a few seedlings may appear, but their establishment is almost nil. Such mass germination, which occurs once in 20 – 30 years, replaces the shrubs that have disappeared over that time, and fills the empty gaps (see Age Groups, Sect. 6.5) (Evenari and Gutterman 1976).

Artemisia monosperma (Asteraceae) is one of the dominant evergreen shrubs found on stable as well as on shifting sands in the deserts of Northern Sinai, the northern and central Negev, and along the Mediterranean shore of Israel. As in *A. sieberi*, the achenes of *A. monosperma* mature and disperse between November and January. The germination of these achenes is much faster (10 days) and is absolutely dependent on light, irrespective of temperature (Koller et al. 1964). In *A. sieberi* the mucilaginous achenes adhere to the soil surface after they have dispersed and are wetted by rain or dew. In both species the times of their maturation, dispersal and germination are very close when germination takes place during the year of maturation. These two factors reduce the risk of losses by consumption by mammals and ants.

3.6 Heterocarpy and Species Survival

Heterocarpy. 'The individual plant produces two kinds of diaspores, one of which is generally telechorous while the other is topochorous' (Zohary 1962). (Telechorous = diaspores capable of wide dispersal; topochorous = diaspores with limitations of transportation). This has been found in disturbed habitats (Harper 1977), deserts and arid habitats (Zohary 1937).

Amongst the vegetation of Mediterranean grove, 82% of plants have telechorous diaspores and only 18% are topochorous. The opposite occurs in Saharo-Arabian Hammada vegetation where 14.5% are telechorous and 85.5% are topochorous (Zohary 1962); 21.6% of the species of the flora of Namaqualand are atelechoric. This includes 30.2% of the geophytes and 28.5% of the annuals (van Rooyen et al. 1990). Among the local flora of Israel, more than 100 species are heterocarpic. This is most common in the families Asteraceae (especially Liguliflorae), Umbelliferae (e.g. Daucinae), Cruciferae (e.g. Brassicinae) and Poaceae (e.g. Hordeae) (Zohary 1962).

The heterocarpic species of the most extreme desert habitats of the central Negev, Judean desert, and others, are the amphycarpic *Gymnarrhena micrantha* (Koller and Roth 1964). Their aerial achenes are telechoric and small, with a relatively large pappus, and they are dispersed by wind. The subterranean achenes are much larger (Fig. 14) and germinate in situ. In *Emex spinosa*, another amphycarpic desert species (Evenari et al. 1977), the aerial propagules are telechoric and have a more 'opportunistic strategy', with germination inhibitors, but higher percentages of germination. The subterranean propagules have a more 'cautious strategy': their germination is more controlled, delayed and spread over time (Chaps. 2, 5, 6).

There is an advantage of survival to species which delay dispersal of their seeds (as mentioned above) as well as the germination of at least some of their seeds (Venable and Lawlor 1980). In North America, the ecological importance of achene dimorphism, dispersal and germination has been studied in arid zones, for example, in *Heterotheca latifolia* (Venable 1985; Venable and Levin 1985) and in *Salicornica europeae* (Ungar 1979).

In polymorphic diaspores of *Dimorphotheca sinuata*, *D. polyptera* and *Ursinia cakilefolia*, collected from populations in the Namaqualand Broken Veld near Springbok, South Africa – 162 mm mainly winter rainfall – morphological as well as germination differences were observed (Beneke 1991; Beneke et al. 1992).

3.7 Conclusions

Under pressure from seed predation and unpredictable environmental conditions, one of the most important questions to be answered is: What proportion of the seeds produced in the Negev desert remains vital for the following rainy season or seasons after the dry hot summer? This relates to plants whose seeds are dispersed after maturation in comparison with plants whose seeds are not dispersed after maturation but are dispersed by the rains of the next season or seasons. Seeds that remain protected by the dead mother plant germinate in situ or are dispersed later. It appears that mechanisms of seed dispersal by rain are more effective under more extreme conditions than immediate dispersal at the beginning of summer. In some of the hottest and most extreme habitats of the Israeli deserts, an ombrohydrochoric species such is *Mesembryanthemum nodiflorum* represents almost 100% of the total number of plants (Gutterman 1990b). The fitness of these species to the local patch habitats is defined by the survival and reproduction. On the other hand, this species is not the most common species of the annual vegetation of the Negev desert. One of the most common species in a wide range of habitats is not *Blepharis*, with the most sophisticated dispersal mechanisms (Sect. 3.3.3.4), nor the amphicarpic or heterocarpic species (Sect. 3.6), but *Schismus arabicus* with its very small grains ('dust seeds') which are dispersed in very large numbers at the beginning of summer. This plant has an opportunistic strategy of seed dispersal and germination (Sect. 3.2.1.1; Chaps. 4, 5, 6). One of the most common perennial plants on the Irano-Turanian hill slopes is *Artemisia sieberi* whose small, mucilaginous seeds are dispersed in winter. Mass germination occurs once in many years and fills 'empty gaps' (Sect. 3.5; Chaps. 5, 6). This leads to the other question: Is there a correlation between seed size, dispersal, germination mechanisms and the survival of populations of annual or perennial species in any particular desert habitat?

There is no pronounced difference in the mechanisms of seed dispersal of annuals and perennials. This is despite the fact that seed germination and the survival of annual seedlings are the most critical events of the life cycle, whereas survival of perennials during summer is more critical. Germination is important to fill 'empty gaps'.

The dynamics of seed dispersal around the year can be summarized as follows (Fig. 52):

Maturation and dispersal at the beginning or during summer. In the majority of annuals, hemicryptophytes and geophytes in the Negev, the seeds mature and are dispersed at the beginning of the summer. The seeds of perennials, such as *Zygophyllum dumosum* and species of *Helianthemum*, mature during July and August. The seeds of common annual and perennial species of Chenopodiaceae mature during July to October, or even November.

During summer in the Negev most seed dispersal is of dust-like, plumed, balloon or winged seeds, by wind (Fig. 53). Ants, mammals and birds, which collect great numbers of seeds are, however, involved only in the dispersal of

Conclusions

the seeds of a minority of species. There are large fluctuations in seed numbers during summer, and seed consumption is mainly by ants. Nearly all the plant species whose seeds are dispersed during summer have very small seeds produced in large numbers, which spreads the risk. There are also a few exceptions, such as *Aegilops* and *Hordeum* spp., with larger, well-protected caryopses. In Namaqualand, 66.3% of the plant species are Anemochorous (dispersed by wind) and only 14.8 are Zoochorous (van Rooyen et al. 1990).

Maturation at the beginning of summer and dispersal by rain in winter. In winter annual plants whose seeds are dispersed by rain, maturation takes place at the beginning of summer. The seeds are protected throughout the summer by the dead parts of the mother plants. At least some of them are dispersed in the rains of the next winter and more in following winters. Seeds are dispersed mainly in December or January when most rain usually falls (Fig. 52). The seeds of some species are dispersed during the following winter and adhere to the soil surface until they germinate, sometimes a number of years later. These seeds exchange one shelter (the capsule of the dead mother plant) for another − the soil crust. This occurs in species of *Mesembryanthemum, Aizoon*, etc. Seeds of plants, such as *Blepharis* spp. and *Asteriscus pygmaeus*, are dispersed when rain falls over a period of many years, a few of them each winter (Tables 39, 40).

Seed surface size and dispersal strategy by rain. Immediately after being dispersed very small seeds adhere to the wet soil and remain as a part of the crust (S) until they germinate (Gutterman 1980/81b; Gutterman 1983) (Table 39). These seeds, as already mentioned, change one shelter (the dry fruit) for another shelter (the soil crust) during the time of rain when seed predators are least active. However, a few hours after the rain, ants are active and collect the seeds that were dispersed by the rain. There are still many seeds that escape because within a short time they adhere to the soil crust and it is difficult to find or separate them from the soil. Very small seeds with a pappus are dispersed by wind (W) after being released from the mother plant by rain. In some species, such as *Gymnarrhena micrantha* and *Asteriscus pygmaeus*, the achenes become stuck to the soil surface by the pappus during subsequent rain (Table 39). Very small seeds which develop a mucilaginous layer (M) on the seed coat after wetting, adhere to the soil surface until they germinate. This mucilaginous contact also acts as a counterforce for the radicle to protrude into the soil (Table 39). The small (1 − 2 mm) mucilaginous seeds of *Anastatica hierochuntica* (Fig. 79), *Reboudia pinnata, Carrichtera annua* (Figs. 131 − 133), *Plantago coronopus* (Figs. 72 − 74) and other species are also dispersed by runoff water. They are transported for a short distance and germinate along the runnels of slopes or in depressions. Thus, even typical atelechoric species are capable of conquering new ecological niches far away from their previous locations and develop new patches (Gutterman 1990b). The relatively large seeds of *Blepharis* spp. (average $6 \times 4 \times 1$ mm) are shot out by the force of the capsule exploding after wetting (Figs. 99, 100), and adhere to the wet soil surface by means of multicellular hairs which develop a mucilaginous layer after wetting (Figs. 103–109). They are dispersed and germinate during the same rain event (Fig. 109). The germination of seeds that are covered by·a mucilaginous layer and dispersed by floods or runoff water is inhibited (Figs. 110, 111) until the excess water disappears (Table 39) (Gutterman et al. 1967; Witztum et al. 1969; Gutterman 1982b). The mucilage-covered seeds may also be dispersed to new areas when they adhere to the legs and beaks of birds or the legs and bodies of mammals passing through the area during the period of rain. Murbeck (1919/20) found that 11% of the North African flora was composed of species with mucilaginous seed coverings; while 20% of the desert flora exhibits myxospermy as compared with only 3% in Scandinavia (van der Pijl 1982). Among the flora of the Namaqualand desert of South Africa, 11.4% of the species are myxospermic (Van Rooyen et al. 1990).

Synaptospermy and desert habitats. Approximately 40 plant species of Israel whose seeds are dispersed by rain are synaptospermic, at least during the first summer after seed maturation. The majority of these are desert species or plants from the more arid areas of the Mediterranean zone. In the arid regions of North Africa, Murbeck (1919/20) found limited seed separation (synaptospermy) in 140 species, of which 100 had entered the Sahara Region. Zohary (1937, 1962) found that 2% of the Mediterranean maquis of Israel consisted of species with synaptospermy, but 19% of the Saharo-Arabian Hammada vegetation of the Negev desert. In southern France, on the other hand, Muller-Schneider (1967) found only 2.5% synaptospermic species and 0% in Fenno-Scandia (van der Pijl 1982). In the Namaqualand desert of South Africa, 7.6% of the flora, but 13.6% of the annual species, are synaptospermic (Van Rooyen et al. 1990).

Telechory and topochory in mesic or arid vegetation. Zohary (1962) compared species with telechorous dispersal mechanisms and species with topochorous (short distance) dispersal from two floral regions: the mesic Mediterranean maquis vegetation of the north of Israel and the arid Saharo-Arabian Hammada vegetation of the Negev. The percentage of plant species with telechoric mechanisms in the mesic region was 82% but, in the arid habitats, only 14.5%. The topochoric plant species of the mesic vegetation composed only 18%, however, in comparison with 85.5% in the arid habitats. The ecological importance of desert annuals to germinate season after season in the same depressions where water accumulates is obvious. In these deserts patchy distribution of local ecogenotypes possibly develops with time. Thus the fitness increases when the gene pool decreases, unless at least three factors counteract to lower it: (1) the long-distance pollination; (2) long-distance seed dispersal; (3) the longer-lived 'old' seeds of the age structure of the seed bank which may germinate together with the 'new' seeds (Went and Munz 1949; Went 1961, 1969; Roberts and Feast 1973; Kivilaan and Bandurski 1981; Brown and Venable 1986; Gutterman 1990a,b).

Conclusions

Seed maturation in summer and germination in situ in winter. The subterranean 'seeds' of amphicarpic annual plant species, such as *Emex spinosa* and *Gymnarrhena micrantha*, do not disperse but germinate in situ where they mature. These plants have a double strategy for seed dispersal. The subterranean seeds remain protected by the dead mother plant from the time of maturation until germination in the same depression, while the aerials are dispersed to a long distance (see below and Sects. 5.1.3 and 6.2.1).

Maturation and dispersal in winter. The seeds of the perennial *Artemisia sieberi*, one of the most common shrubs of the Negev, mature and their mucilaginous achenes are dispersed during December and January. Mass germination of this plant occurs once in many years, which causes age groups to form, as also in *Zygophyllum dumosum* (Evenari et al. 1971, 1982). The yearly fluctuations in seed dispersal and accumulation in the seed bank are summarized and show that seeds of different species are dispersed during different seasons of the year (Fig. 52).

A most pronounced characteristic is the small size of the seeds of nearly all plant species of the Negev desert. Under pressure from seed consumption, an important strategy for plant survival is the production of many very small seeds (0.5–0.7 mm, as in *Schismus arabicus*) (Sect. 3.2.1.1). The few species that produce larger seeds, produce relatively small numbers but they are well protected. Examples are afforded by *Blepharis* spp. (average $6 \times 4 \times 1$ mm), whose seeds are released and germinate during the same rain event, the subterranean achenes of *Gymnarrhena micrantha*, and the subterranean propagules of *Emex spinosa*, which germinate in situ.

It is very difficult to reach a general conclusion as to whether there are strategies of seed dispersal which are typical of desert plants and do not exist in plants that inhabit more humid areas. The mechanism of seed dispersal and germination of species such as *Blepharis* spp. is one of the few examples of a very sophisticated adaptation to completely unpredictable desert conditions (Gutterman et al. 1967).

In Israeli desert areas, where depressions are favourable micro-habitats, atelechoric species, such as *Mesembryanthemum nodiflorum* (Gutterman 1980/81a), *Trigonella stellata* (Evenari and Gutterman 1976) or *Plantago coronopus* (Gutterman et al. 1990), whose seeds are rain dispersed, inhabit patches of diggings and depressions together with other species. The atelechoric mechanisms not only give them a better chance to germinate in the same favourable micro-habitat in which their mother plant survived successfully and produced seeds, but also encourage the appearance of local ecotypes which are especially well adapted to these particular desert conditions. On the other hand, for plants to be in very small groups separated from the other populations of the species, reduces the gene bank which may decrease survival rates. This may be overcome by at least three main mechanisms: (1) long-distance pollination, (2) germination of the longer-living seeds of the seed bank; (3) a double strategy, such as that of amphicarpic plants or plants with dimorphic or heteromorphic seeds. In these some of the seeds germinate in situ or at a short distance from the mother plant, and others are dispersed to a long distance. *Ifloga spicata*, *I. rueppellu* and *Filago desertorum* have two strategies of dispersal: achenes with a pappus for long-distance dispersal by wind and atelechoric achenes lacking a pappus. They enjoy a high survival rate in patchy micro-habitats. Seeds that germinate in situ have more 'cautious' strategies for germination and survival, whereas telechoric seeds have more 'opportunistic' strategies (e.g. *Emex spinosa*) (Evenari et al. 1977). A double strategy existing in one species is an optimal model for survival in a heterogeneous environment, as shown by Hamilton and May (1977), Comins et al. (1980), Motro (1982), and Venable (1985).

4 Storage Conditions Affecting the Germinability of Seeds in the Seed Bank

4.1 Introduction

In addition to the genotypic inheritance affecting seed germination of each plant species, and the environmental and maternal influences of each seed during maturation (Chap. 2), there are also important influences during the period of storage on the germinability of each seed (the period between seed maturation and seed germination). These are dependent upon the location of each seed in time and space, in its particular micro-habitat, during the time of storage. The position of the seeds in the dead mother plants, on the soil surface, or below the soil surface, together with the influences of daily and seasonal environmental factors, may have an affect on seed germination.

Important queries about possible influences on seed germination during the period of seed storage are concerned with: (1) The location and dynamics of change in each location in a particular micro-habitat, such as on the soil surface or at different depths in the soil, and the short- and long-term dynamics of soil turnover (Fig. 113 a, b, c). (2) The possible changes in the germinability of each seed depending upon the environmental factors and location of micro-habitat from maturation during the hot and dry summer until germination during the following rainy season. In the Negev, the majority of seeds mature at the end of the growing season (at the beginning of summer) (Sects. 1.2.3, 1.4, 3.1; Fig. 52). (3) Changes in the viability of seeds over the short term (one year or less) and long term (many years).

In conclusion, a final question: What influences during the period of storage affect 'readiness for germination' of seeds in a particular rain event?

4.1.1 Seed Bank Location of Some Desert Plant Species

In areas which are covered with the typical soil crust, the seed banks can be located as follows (Figs. 113 a, b, c, 114): (1) As dispersal units fruits containing one or more seeds, as inflorescences (synaptospermy) situated on or below the soil surface or, in the case of species whose seeds are dispersed by rain, on the dead mother plant (Murbeck 1919/20; Zohary 1962; Gutterman 1990 b). (2) The subterranean propagules or achenes of the amphicarpic species, such as *Emex spinosa* (Evenari et al. 1977) and *Gymnarrhena micrantha* (Koller and Roth 1964) are situated a few millimeters to 1 or 2 cm below the soil surface

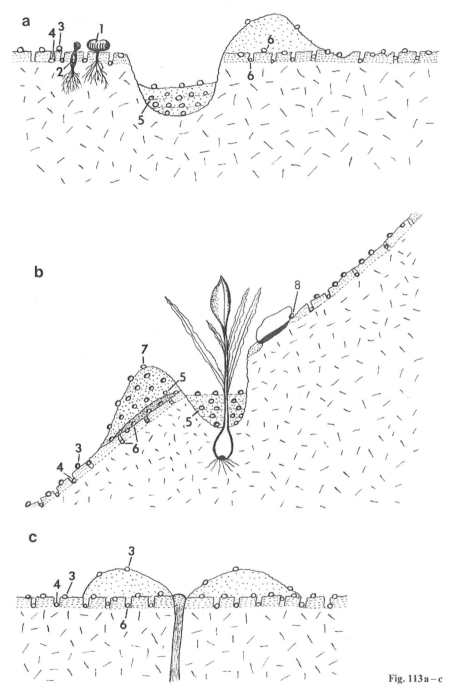

Fig. 113a–c

Introduction 147

(basicarpy) (Zohary 1962). (3) On the soil surface to which mucilaginous seeds adhere (Gutterman et al. 1967) (myxospermy) (Zohary 1937, 1962), or very small seeds which adhere to the soil crust by dew or rain shortly after being dispersed (Gutterman 1980/81 a). (4) Very small seeds which are dispersed by wind or floods are situated in cracks in the soil crust (Loria and Noy-Meir 1979/80). (5) Other seeds accumulate as a result of wind or floods near or beneath stones, below shrubs, in depressions and porcupine diggings, and are covered by sediment (Gutterman 1982c, 1988c; Gutterman et al. 1990). (6) Seeds situated above or near the soil surface may be covered by mounds formed by animal activity, as well as sediments accumulated as a result of wind and floods. In many desert habitats, the majority of the seeds of the seed bank are situated either above the soil surface or at the depth of 1 or 2 cm below it (Sect. 3.2.4). In the Sonoran Desert, 89% of the seeds are located in the upper 2 cm of the soil (Childs and Goodall 1973).

4.1.2 Soil and Seed Turnover

The position of seeds that are situated on or near the soil surface can be changed by wind, flood and animal activity. In some areas intensive porcupine or isopod digging overturns and changes the position of seeds in the soil (Fig. 113 a, b, c): (1) When porcupines dig for subterranean storage organs (corms and bulbs) or subterranean parts of leaves and stems of geophytes and hemicryptophytes a hole, 20 – 30 cm in depth, is left. In some habitats this hole will be filled during the course of 6 to 15 – 20 years. During this time seeds, dispersal units, organic matter and soil particles accumulate in it in summer, as a result of wind. Seeds that are carried in by water, and soil particles freed

---

Fig. 113 a. A porcupine digging where the geophyte (e.g. *Bellevalia desertorum*) was consumed the first time and destroyed (Gutterman 1982c, 1988a). Possible location of seeds during the time of storage after maturation and before germination (Sect. 3.2.2): (*1*) a dead mother plant of a species which disperses its seeds by rain (e.g. *Asteriscus pygmaeus* (cross section). The main seed bank is situated in the inflorescence of the dead mother plant above the soil surface (Sect. 3.3.3.2); (*2*) an amphicarpic plant with basicarpic subterranean achenes that have matured below the soil surface and are germinating in situ (cross section) (Sect. 3.4); (*3*) myxospermic seeds which adhere to the soil surface by their mucilage, or very small seeds that adhere to the soil surface during dispersal by rain when the soil surface is wet; (*4*) very small seeds that fall into cracks in the soil crust (Sect. 3.1); (*5*) seeds that have been trapped in porcupine diggings and covered by layers of sediment each year (Sect. 3.2.2); (*6*) seeds situated in cracks and on the soil surface which have been covered by the mound of soil which the porcupine excavated whilst digging. **b** Porcupine digging in which there is a plant that is partly consumed (e.g. *Tulipa systola* on the hillslope, or *Bellevalia eigii* in the wadi) (Gutterman 1982c, 1988a). In this system, after full cover-over of the digging the seeds (*5*) were buried again in the new hillock after re-excavation of the porcupine digging. This will cover the remainder of the old hillock; (*1*) to (*6*) − as in **a**; (*7*) seeds buried in the previous mound resurface after soil erosion; (*8*) seeds situated beneath a stone. **c** Isopod families dig nests and carry out the faeces containing soil which accumulate in a mound surrounding the nest (Shachak and Brand 1988, 1991). This mound could cover seeds on the soil surface and in cracks and disappear, usually after 1 year

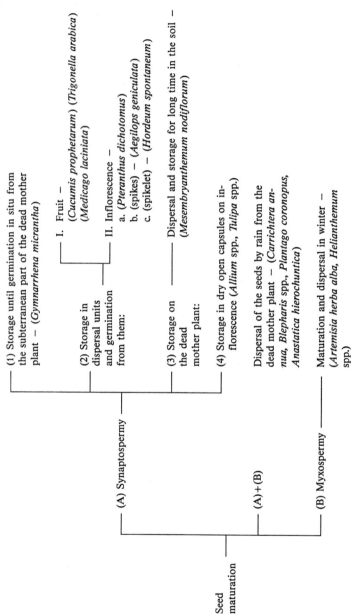

Fig. 114. Storage conditions of some common plants of the Negev highlands. (*A*) synaptospermy (Murbeck 1919/20; Zohary 1962); (*B*) myxospermy (Zohary 1937, 1962) mucilaginous seed coats

Introduction 149

by erosion during floods, accumulate in the digging in winter. Some of the seeds germinate, and others are buried at the bottom of the digging. Every year, more seeds are left buried at higher levels (Fig. 113a, b). As porcupines, isopods and other animals dig more in some areas than others, the soil crust is damaged and more soil particles are freed. These are carried and trapped in depressions and porcupine diggings. This reduces the time taken to fill the diggings. When a hole is covered over and geophytes within become again attractive to porcupines, they are once again consumed. In the process, the seeds and soil particles that had accumulated over 6 to 15–20 years are removed, creating a mound near the digging (Gutterman 1982c, 1988a, 1989a; Gutterman et al. 1990). When this mound is eroded by wind and rain, the seeds are released from the soil and dispersed nearby. Mounds also cover seeds which are situated on or near the soil surface (Fig. 113a, b). (2) Seeds adhering to the soil surface or accumulating in soil cracks may also be covered by soil or by aggregates from the excretion and defaecation of the isopods (Fig. 113c) (Yair and Shachak 1987; Shachak and Brand 1988, 1991). The activities of ants, birds and other animals also contribute to changes in the soil surface by increasing free soil particles. (3) Free soil particles accumulate in depressions or form mounds surrounding shrubs and cover seeds that have accumulated below these shrubs. (4) Floods change the soil surface by erosion and submergence. (5) Seeds on the soil surface or in cracks of the soil crust that are covered by the mound formed when porcupines excavate, or surrounding isopod and ant nests, appear on the soil surface after these mounds have been eroded by wind and floods.

4.1.3 Environmental Factors and Changes at Different Soil Depths

4.1.3.1 Temperature

Depending on their position in or on the soil, seeds are exposed to different daily and seasonal maximum and minimum temperatures. The deeper they are, the smaller are the fluctuations in temperatures between day and night, as well as seasonally. Temperatures on the soil surface are extreme in all seasons but especially so in summer when the large areas between the small shrubs are bare. Daily fluctuation can then be from 10 °C to above 50 °C. At a depth of about 20–30 cm, however, daily fluctuations almost disappear and the temperature is about the daily average of the surface minimum and maximum (Shachak 1975). In the autumn soil surface temperatures higher than 55 °C have been measured, and light intensity of 1450 μmol m^{-2} s^{-1} at noon (Gutterman and Boeken 1988). Similar patterns were observed on 16–22 January 1939 by West (1952) in Griffit, New South Wales, Australia. Typical fluctuation patterns of the temperature of bare soil in midsummer were found to be ca. 60/15 °C in Northern west Australia (Quinlivan 1965; Cloudsley-Thompson 1991).

4.1.3.2 Light

Germination of the seeds of *Artemisia monosperma* (Koller et al. 1964), *A. sieberi, Portulaca oleracea* (Gutterman 1974), *Carrichtera annua* (Gutterman 1990a), *Lactuca serriola* (Gutterman 1992c), etc., is promoted by light, so they do not germinate poorly when buried deeply in soil. Only after the soil turnover has exposed them to light, will they germinate. Light penetration into the soil is very limited and depends upon the soil colour, structure and crust. Enough light for the germination of light-sensitive seeds cannot penetrate most soils for more than 2 mm. Moreover, the visible wavelength of the light increases with soil depth until only Far Red penetrates (Woolley and Stoller 1978). This inhibits the germination of light-sensitive seeds (Mayer and Poljakoff-Mayber 1982; Bewley and Black 1982). Achenes of *Artemisia monosperma* need to be deep enough in the sand to reach moisture but not too deep because they require light for germination. Any wavelength of visible light increases the germination of *A. monosperma* in comparison to dark (Koller et al. 1964).

The seeds of some desert species, however, are inhibited by light. They need to be covered by soil before they are able to germinate. Examples are afforded by the seeds of *Pancratium maritimum* (Sect. 3.2.1.1) and many others (Sect. 5.5.4.2) which have to be covered by sand before germination (Keren and Evenari 1974), and of *Calligonum comosum* (Koller 1956). Interactions between light and temperature often have an influence on seed germination (Bewley and Black 1982). Germination of *Atriplex dimorphostegia* is inhibited by light at 20 °C and 26 °C (Koller 1957). Seeds of *Hyoscyamus desertorum* germinate at temperatures of 25 – 30 °C in continuous light, but not at 10 – 20 °C. However, they will germinate at 15 – 20 °C in the dark if they are illuminated for 10 min per day (Roth-Bejerano et al. 1971).

4.1.3.3 Wetting and Drying

Seeds that adhere to the soil surface are wetted by dew at night (Sect. 1.2.2), and subsequent drying during the day about 190 times per year, in addition to some 10 – 30 days of rain, near Sede Boker (Zangvil and Druian 1983). Seeds that lie deeper in the soil are only wetted by rain (Evenari and Gutterman 1976; Evenari et al. 1982). Wetting or increase in R. H. may be important for the longevity of seeds. It enables damage that occurs during storage to be repaired (Karssen et al. 1989). The high salinity that accumulates in summer on the surface of desert soils, together with dew, may increase germination rates, uniformity (Heydecker et al. 1973) and possibly the number of seeds that will be 'ready for germination' during the following rainy season. Seeds on the soil surface are exposed to extreme conditions. The dynamics of the soil affects dramatically the environmental influences to which they are exposed. Germinability is affected during storage and this has an influence on the number of seeds that are 'ready for germination' during following rain events (see below). 'Readiness for germination' is, therefore, dependent on the genotypic

Seed Internal and Environmental Factors During Storage Affecting Germination 151

inheritance of the species, the history of each seed (possibly including the position of the inflorescence on the mother plant or the position of the seed in the inflorescence or in the fruit). Environmental factors, such as day length, temperature, age of mother plant at seed maturation, water stress, and so on, during maturation, also have an affect (Chap. 2). All of these, as well as dispersal (Chap. 3), storage conditions, and conditions during germination, may affect the germinability of a seed (Chap. 5).

4.2 Seed Internal and Environmental Factors During Storage Affecting Germination

4.2.1 Seed Coat Effects

During the time of storage, the seed coat is exposed to various environmental factors that affect germinability and viability. At least seven factors are involved in seed-coat imposed dormancy (Ballard 1973; Rolston 1978; Werker 1980/81; Bewley and Black 1982; Egley and Duke 1985), and protection against pathogens and seed eaters. These are: (1) seed coat impermeability prevents water from reaching the embryo; (2) seed coats prevent the dilution of germination inhibitors around the embryo; (3) seed coats limit respiratory gas exchange; (4) seed coats prevent light from reaching the embryo; (5) the seed coat acts as a mechanical barrier against embryo expansion; (5) seed coat structure is resistant against pathogens; (7) seed coats may contain poisons that deter seed eaters. Changes of the seed coat and other factors situated in the embryo and/or in the endosperm during storage may lead to changes in 'readiness to germinate' as well as in viability.

Species with impermeable seeds are most numerous among the Leguminosae, but representatives with hard seeds can also be found in the Malvaceae, Chenopodiaceae, Geraniaceae, Convallariaceae, Convolvulaceae, Solanaceae, Cannaceae, Poaceae and Liliaceae. The impermeability of the coat of hard seeds results from the pressures of impregnation of different hydrophobic substances in the cell walls (Kolattukudy 1980a, b, 1981, 1984; Werker 1980/81; Egley et al. 1985; Egley 1989). Openings in the hard seed coats are the hilum valve (Hyde 1954) micropyle and chalaza (Marchaim et al. 1974). The strophiale or lens zone is weak in the hard seeds and environmental factors can induce opening and water penetration (Hagon and Ballard 1970).

4.2.1.1 Changes Occurring in 'Hard Seeds' During the Time Between Maturation and Germination

Seeds of desert plants, such as *Ononis sicula*, are dispersed from the dry pods after maturation at the beginning of the summer. They then roll along the soil surface (Sect. 3.2.1; Fig. 53). The day length and plant age during maturation were found to affect the proportion of yellow/green and brown seeds in the

152   Storage Conditions Affecting the Germinability of Seeds in the Seed Bank

seed population (Sect. 2.4.2.1; Figs. 21–28) (Gutterman 1973). The yellow seeds which mature under long days become 'hard seeds'. There are at least three methods of bringing them to germination during storage: (1) seed coat scarification; (2) softening by the hilum valve mechanism; (3) softening by temperature fluctuation.

4.2.1.2 Scarification of the Seed Coat

Scarification by wind and soil particles, or damage to the seed coat by rolling along the soil surface are more effective on the green seeds of *Ononis sicula* than on yellow seeds. Yellow seeds have the most developed seed coat (Sect. 2.4.2.1; Figs. 24, 25, 28) and remain as hard seeds for a longer time. The process is much shorter in green seeds with less developed seed coats (Fig. 27), which either matured under short days or on senescent plants under long days. Brown seeds with the least developed seed coats imbibe water immediately upon wetting. Following maturation, they may germinate during the first rains of the season. The brown seeds that do not germinate may die, however, because the undeveloped seed coat does not protect the embryo well. They can, therefore, be damaged by the activity of micro-organisms in the soil. The day length response of the seeds of *Trigonella arabica* is similar (Figs. 29, 30) to that of *Ononis sicula* (Gutterman 1973, 1978a, 1992b) (Sect. 2.4.2.1), but, in this species, after maturation the seeds remain in the unopened pod which contains germination inhibitors (Sect. 5.2.2.1) (Koller 1954).

Day length in late spring (the growing season), and plant age, were found to affect the proportion of hard seeds of subterranean clover (*Trifolium subterraneum*) under natural conditions, 50 km east of Perth, West Australia (Quinlivan 1965) (Sect. 2.4.6.1). Cutin composition may be affected by environmental factors and thereby affect seed coat permeability (Kolattukudy 1980a).

On the sandy Mediterranean seashore of Israel, hard seeds of *Retama raetam* that have passed through the digestive system of animals and have been removed from their faeces, germinate to a much higher percentage (50%) than seeds collected in the same area from pods (2%) (see Sect. 3.2.3; Fig. 75). In Australia, similar results were reported by Leigh and Nobel (1972) on the germination of *Nitraria billardieri* seeds collected from the faecal droppings of emus. This was much higher than that of seeds collected from shrubs (50% to 3%) (Sect. 3.2.3) (Noble 1975b).

Unintested seeds of *Acacia raddiana* in the Negev, taken from gazelle droppings, germinated after 10 days to 21% when the control germinated to only 4%. In this population 72% of the seeds were infested by seed beetles (Bruchidae) (Sect. 3.2.3) (Halevy 1974).

Many trees and shrubs in the Mojave and Colorado deserts of North America, such as *Dalea spinosa*, *Olneya tesota* and *Cercidium aculeatum*, are restricted to the bottoms of wadis. These trees have extremely hard seeds. During a flood after heavy rain, seeds are washed down together with sand and gravel and their coats are scarified. This allows the embryo to break through after absorbing water (Sect. 5.1.1) (Went 1953).

4.2.1.3 The Role of the Hilum Valve Depending on Relative Humidity During Storage

According to the fascinating study of Hyde (1954), there is a connection between the changes in relative humidity during storage and the ability of clover (*Trifolium repens, T. pratense*) and lupin (*Lupinus arboreus*) seeds to germinate. In white clover (*T. repens*), Hyde found that seeds reduce their water content to 25% at the period of maturation and then more slowly to 14%. At approximately this stage, seed coats become water-impermeable (hard seeds). This phenomenon has also been found in other species with hard seeds (Egley 1989). From this stage, only water vapour diffusion through the hilum takes place. If the relative humidity (R.H.) decreases from 70% to 30%, the water content of the seed is reduced to about 8%. When the R.H. of the atmosphere around the seed is about 0%, the water content of the seeds is reduced to about 1%. During 7 days in 70% R.H. the water content of the scarified seeds increases to 14%, but there is no change in the water content of the non-scarified seeds (Fig. 115). When the hilum was blocked, the water content did not change from 11% which it had been at the beginning of the experiment

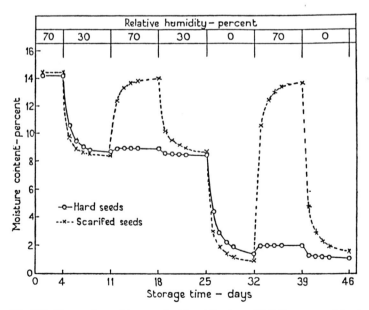

Fig. 115. Changes in moisture content in white clover (*Trifolium repens*) hard or scarified seeds transferred once a week successively to chambers of different relative humidity (70%, 30%, 0%) (Hyde 1954)

154 Storage Conditions Affecting the Germinability of Seeds in the Seed Bank

Table 49. Moisture content (% of dry weight) of seed samples of Red clover after successive periods of storage under different humidity conditions. (After Hyde 1954)

Air	Seed moisture content		
Relative humidity %	Hard seeds	Hard seeds with hilum blocked	Scarified seeds
65	11.9	11.8	12.0
50	9.5	11.8	9.4
33	7.8	11.6	7.4
70	8.2	11.8	12.5
0	1.2	11.5	0.9
70	1.9	11.7	12.4

(Table 49; Fig. 116a, b). When seeds contain more than 14% water they germinate when immersed in water. When the water content decreases from 14% to 8% there is a higher percentage of 'hard seeds' in the population and a lower percentage of germination (Figs. 117, 118). At a water content of 8%, seeds are completely hard and none germinates in contact with water (Figs. 115, 118). When seeds are immersed in water at this stage the coats swell to a degree which closes the hilum valve so that water is not able to reach the embryo (Fig. 116a, b). Germination is, therefore, prevented. Such seeds are called 'hard seeds'.

Hyde (1954) also found that water vapour could enter through the hilum valve at all stages. A gradual increase in relative humidity increases the volume of the embryo. Water vapour is absorbed into the embryo which increases in volume to such a degree that the hilum valve opens sufficiently for liquid water to penetrate (Fig. 116a, b). This occurs in spite of the fact that when the seed is soaked in water, swelling of the coat causes a degree of closure of the hilum valve. There is a close correlation between the germination of white clover and storage at different levels of relative humidity (Fig. 118). It is possible that such mechanisms may also act in deserts. In micro-habitats, where the soil retains high relative humidity between one rain event and another, the volume of the embryo may increase to such an extent as to allow water to penetrate the hilum valve of the seed. In some years large numbers of *Ononis sicula, Trigonella arabica*, and other species that have hard seeds appear. In other years, far fewer of these species are to be seen.

Medicago laciniata is a Saharo-Arabian desert annual also inhabiting the Negev deserts. After 2 months of constant relative humidity ranging from 13% to 83.5%, the percentage of seeds that swell during imbibition in water increased progressively to the increase of the relative humidity during storage, up to a maximum of 55% (Koller 1972).

In *Lathyrus aphaca*, which also inhabits some deserts of Israel in the Irano-Turanian region, the 'lag phase' before seed swelling was longer after storage in low relative humidity; only 12–25% of swelling per day. In comparison, the swelling of seeds stored above relative humidity of 80.5% was 91% per day.

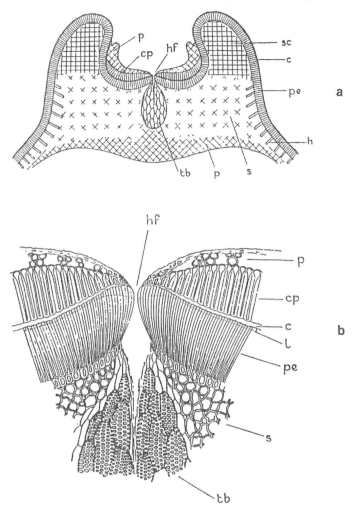

Fig. 116. **a** Diagrammatic illustration of the tree lupin seed showing a transmedian section of the hilum and adjacent tissues; *p* parenchyma; *cp* counter-palisade; *hf* hilar fissure; *sc* sclerenchyma; *c* cuticle; *pe* palisade epidermis; *h* sub-epidermal layer of 'hour glass' or 'pillar' cells; *s* stellate cells; *tb* tracheid bar (Hyde 1954). **b** Higher magnification of the hilar fissure area

However, the seeds of this plant from all the storage treatments were fully permeable at the end of the experiment (Koller 1972).

In seeds of *L. hierosolymitanus*, a Mediterranean plant also inhabiting the Negev and Judean deserts, storage in relative humidity above 49% increased swelling from 11–14% per day to 76–78% per day after storage in 80.5–83% relative humidity (Koller 1972).

The relative humidity during storage may affect the spread of germination in time. Seeds of species inhabiting the more extreme desert conditions, such

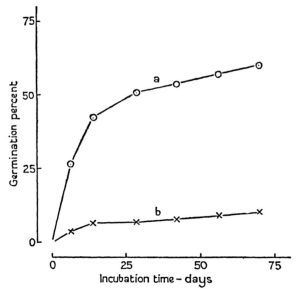

Fig. 117. Rate of germination of 'hard seeds' of white clover (*Trifolium repens*) previously stored under different conditions; **a** seed stored under conditions of gradually increasing humidity; **b** seed stored at constant high humidity. (Hyde 1954)

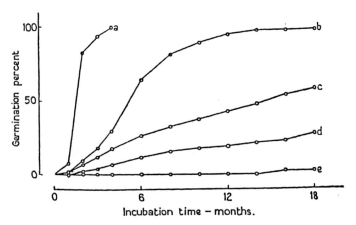

Fig. 118. Rate of germination of 'hard seeds' of white clover (*Trifolium repens*) ripened at the following levels of relative humidity: *a* 70; *b* 60; *c* 50; *d* 40; *e* 30%. (Hyde 1954)

as *M. laciniata*, swell the slowest and only a portion of the seeds germinate in comparison to the other two species (Koller 1972).

4.2.1.4 Temperature Fluctuations Affecting Legume Hard Seeds Softening

Quinlivan (1961, 1965, 1966, 1968) found in Australia that subterranean clover (*Trifolium subterraneum*) and blue Lupin (*Lupinus pilosus*) seeds show a high proportion of germination during the winter following a spring season of overgrazing. The proportion of seeds that germinate is lower when they lie in soil covered by plant remains. Here the range of day/night temperature fluctuation is lower. Under laboratory conditions, the greater the daily temperature fluctuation during 6 months of dry storage, the higher the percentage of germination. Almost 100% softening of the hard seeds was observed in seeds stored under 75/15 °C (Fig. 119) (Quinlivan 1966). Similar results have also been observed in Western Australia in *Ornithopus compressus* (Barrett-Lennard and Gladstones 1964). Daily temperature fluctuation has an effect on the split which develops at the strophiolar zone (Hagon and Ballard 1970) as well as on 'impaction' (Hamly 1932). Impaction for 12 h of *Aspalathus linearis* seeds collected at the South-Western Cape of South Africa affected permeability at

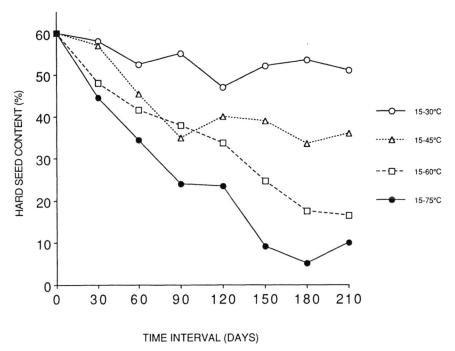

Fig. 119. The rates of softening of Geraldton subterranean clover (*Trifolium subterraneum*) seeds, during 210 days of dry storage with increasing amplitudes of the temperature fluctuations. (After Quinlivan 1966)

158   Storage Conditions Affecting the Germinability of Seeds in the Seed Bank

the lens (= strophiolar zone) and 100% germination resulted (Kelly and van Staden 1987). Squeezing or striking them may also affect 30–90% of the seeds at the strophiolar zone (Ballard 1976) so that water is able to penetrate and cause germination. Water penetration through the strophiolar split depends on mechanisms similar to those described by Hyde (1954) in *Trifolium repens, T. pratense* and *Lupinus arboreus*. It is probable that similar results may be obtained in other species of this family, such as *Ononis sicula* and *Trigonella arabica* which have hard seeds. The lens is the site of initial water entry and regulates the water uptake which increases seed and seedling vigour. This was found in seeds of *Sesbania panicea* collected near Pietermaritzburg, South Africa (Manning and van Staden 1987).

4.2.2 Short, Wet Storage at High Temperatures

Imposed thermo-inhibition or thermodormancy may occur when seeds are wetted by rain in summer, when temperatures are high. It is possible that such mechanisms prevent winter annuals from germinating in summer, or perennials which inhabit areas receiving mainly winter rains, from germinating after summer rain. Under such conditions, there is an increase in the summer-germinating species inhabiting desert areas receiving summer rain (Gutterman 1990a) (Sect. 5.4.4).

4.2.3 The Short-Term Changes of Seeds Influenced by Temperatures During Dry Storage

In *Hordeum spontaneum*, a Mediterranean plant of which southern populations inhabit wadi beds in the Negev highlands near Sede Boker, a period of high temperature is required before the caryopses are 'ready for germination'. In its natural habitat, one caryopsis (spikelet) or dispersal unit becomes entangled and buried to a depth of 2–3 cm below the soil surface at the beginning of summer. Here it is exposed to high temperatures. The caryopses collected after maturation do not germinate, after imbibing water, unless they are exposed to high temperatures (above 35 °C) during summer. After this they become 'ready for germination' by the beginning of the winter. This was observed when caryopses collected from their natural habitat at the end of the summer were soaked. This was not the case with those which were soaked at the beginning of summer.

Freshly harvested dispersal units do not germinate after imbibition, either in the light or dark, in a range of temperatures from 5–35 °C (Fig. 120.I). Nevertheless, the separated embryos do germinate (Fig. 120.II). Even dispersal units with a split in the endosperm close to the embryo are able to germinate (Fig. 120.III). If the split is some distance from the embryo, however, the seed does not germinate (Fig. 120.IV). Embryos still attached to dispersal units germinate when the 'coats' that cover them are removed (Fig. 120.V). Naked

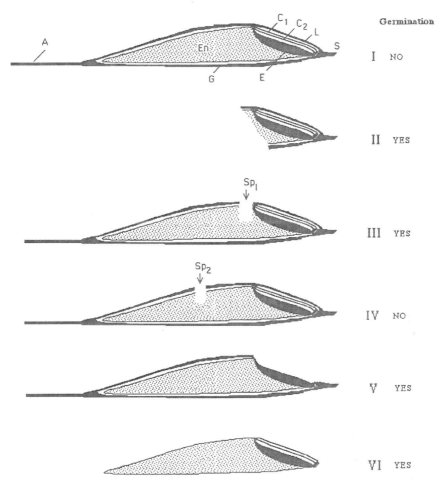

Fig. 120. *I* Schematic drawing of longitudinal section of *Hordeum spontaneum* dispersal unit: *E* embryo of caryopsis; C_1, C_2 the two cover tissues of the embryo; *En* endosperm of caryopsis; *G* glumella; *L* lemma; *S* spikelet stalk; *A* awn. *II* Separated section with the embryo in its covers attached to a small part of the endosperm; *III* split in the endosperm close to the embryo (Sp_1); *IV* split in the endosperm at a distance from the embryo (Sp_2); *V* naked embryo in a dispersal unit; *VI* naked caryopsis with the two covers of the embryo. (Unpubl. data)

caryopses with the two covers to the embryo, also germinate (Fig. 120.VI) (unpubl. data). About 70 days of storage in dry surroundings at 35 °C are needed to overcome afterripening. Dispersal units from the same harvest that had been kept at −35 °C did not germinate either during the experiment or even at the beginning of the following winter. The long period of high temperature needed before grains are 'ready to germinate' is a very important survival mechanism because it prevents germination in the wrong season (unpubl. data). It has been known for some time that storage at high temperatures overcomes the dorman-

160 Storage Conditions Affecting the Germinability of Seeds in the Seed Bank

cy (afterripening) of wild oats (*Avena fatua*) as well as of *A. ludoviciana* (Johnson 1935; Thurston 1951, 1962; Quail and Carter 1969).

Temperature during storage had a considerable influence on seed germination in eight out of nine annual plant species from the Mojave and Sonoran deserts (Capon and Van Asdall 1967). After dry storage in constant temperature (50 °C) for 1 – 5 weeks, very high percentages of seeds of eight species germinated. In contrast, only low percentages (0 – 14%) of untreated seeds of these eight species germinated. Storage at 20 °C for 4 – 8 weeks also results in relatively poor percentages of germination (0 – 16% after 4 weeks; 0 – 30% after 8 weeks). In *Salvia columbariae* and *Geraea canescens*, 80% and 60% germination respectively were obtained after 1 week of storage in dry conditions at 50 °C. In *Lepidium lasiocarpum* and *Eriophyllum wallacei*, too, 90% and 48% germination respectively were obtained after 2 weeks at 50 °C. After 3 weeks, *Sisymbrium altissimum* reached 92% germination (Table 50). Storage at 4 °C gave low percentages of germination, while seeds of nine desert annuals did not germinate after storage at 75 °C (Capon and Van Asdall 1967). Kurtz (1958) found that seeds of saguaro cactus (*Carnegiea gigantea*) survive after 7 days of dry storage at 83 °C.

In some Negev desert plants, such as *Panicum turgidum*, the percentage germination of seeds from dispersal units dried over $CaCl_2$ before sowing increased below 30 °C, but not under low temperatures (5 °C). An increase in seed germination has also been found in the dimorphic dispersal units of *Atriplex dimorphostegia* when dried over $CaCl_2$ before sowing (Koller 1954).

Table 50. Germination (%) of nine species of annual desert plants from the Mojave and Sonoran deserts, dry-stored under constant 20 °C and 50 °C, compared with seeds stored at 4 °C. (After Capon and Van Asdall 1967)

Species	Storage at 4 °C	Storage at 20 °C for 8 weeks	Storage at 50 °C for			
			1 week	2 weeks	3 weeks	5 weeks
Coreopsis bigelovii	9	30	26	16	24	32
Eriophyllum wallacei	5	24	32	48	30	36
Euphorbia polycarpa	1	2	2	10	12	2
Geraea canescens	14	16	60	74	32	48
Salvia columbariae	6	14	80	26	28	16
Lepidium lasiocarpum	0	4	53	90	72	54
Sisymbrium altissimum	10	12	12	13	92	71
Streptanthus arizonicus	0	8	21	23	29	50
Plantago insularis (not leached)	0	0	0	0	0	0
Plantago insularis (leached)	100	–	100	–	–	–

4.2.4 Long-Term Changes During Dry Storage and Viability

The longer the time of storage at high temperatures (40 °C), between 1 and 16 months, the higher the percentage of germination after 24 h in darkness in *Portulaca oleracea* seeds at constant temperatures of 35, 40 and 45 °C (Table 51). After storage at room temperature for about 11 months, however, the percentage of germination, after 8 h imbibition in the dark, was lower than after storage at 40 °C (Table 52). *P. oleracea* seeds can begin to germinate within 3 – 4 h after 16 months dry storage at 40 °C (Table 53) (Gutterman 1969).

As already mentioned (Sect. 2.2.7), the very small seeds of *Mesembryanthemum nodiflorum* kept in their dry capsules under laboratory conditions did

Table 51. Comparison of germination of two *Portulaca oleracea* seed populations: N[a] and LD[b], after 24 h of imbibition in temperatures of 35, 40 and 45 °C, in the dark. Germination experiments were carried out on the dates listed on the table, between 16. 8. 1967 and 5. 11. 1968, i.e. after 1 to 16 months of dry storage at 40 °C. (After Gutterman 1969)

Date of experiment	Temperatures of imbibition and germination (% ± S. E.)					
	35 °C		40 °C		45 °C	
	N	LD	N	LD	N	LD
16. 8. 67	2.5 ± 1.6	0	8.0 ± 1.4	6.0 ± 1.4	8.0 ± 2.8	1.5 ± 0.8
1. 11. 67	20.0 ± 2.3	1.0 ± 0.7	37.0 ± 3.1	7.5 ± 1.5	46.5 ± 6.4	18.5 ± 3.3
17. 12. 67	31.0 ± 4.5	3.5 ± 2.6	37.0 ± 3.9	14.5 ± 3.6	51.5 ± 3.4	18.0 ± 2.1
21. 1. 68	31.0 ± 3.9	2.0 ± 0.9	41.5 ± 1.1	14.0 ± 2.1	53.5 ± 6.6	31.5 ± 4.9
6. 3. 68	37.5 ± 6.1	6.5 ± 2.6	43.0 ± 2.7	20.0 ± 1.3	52.0 ± 1.9	31.5 ± 3.2
28. 4. 68	39.5 ± 4.6	10.5 ± 1.7	55.5 ± 5.5	33.5 ± 5.4	54.5 ± 2.9	36.0 ± 4.3
9. 7. 68	36.5 ± 3.3	7.5 ± 2.4	54.5 ± 3.8	34.5 ± 2.0	56.0 ± 6.1	39.0 ± 7.8
5. 11. 68	40.5 ± 4.7	13.5 ± 3.9	60.0 ± 4.9	44.0 ± 0.9	55.0 ± 2.0	41.0 ± 4.8

[a] From natural habitat in Jerusalem harvested 14. 7. 1967.
[b] Seed population matured under 16 h day length at temperatures of 27/22 °C harvested 11. 7. 1967.

Table 52. *Portulaca oleracea* seed germination (% ± S. E.) after 8 h of imbibition in 40 °C in the dark after about 11 months of dry storage of two populations LD[a] and N[b] at 40 °C or at room temperature. (After Gutterman 1969)

Seed population	Storage conditions	Germination (% ± S. E.) after 8 h of imbibition
LD	40 °C	42.0 ± 4.0
	Room temperature	17.0 ± 2.0
N	40 °C	52.0 ± 3.0
	Room temperature	33.0 ± 3.9

[a] Seed population matured under 16 h day length at temperatures of 27/22 °C harvested 11. 7. 1967.
[b] From natural habitat in Jerusalem harvested 14. 7. 1967.

Table 53. *Portulaca oleracea* seed germination (% ± S.E.) 0 to 48 h of imbibition in continuous light in 40 °C, after about 16 months of dry storage at 40 °C. Experiment carried out on 5. 11. 68 of two populations of seeds: N[a] and LD[b]. (After Gutterman 1969)

Population	Germination (% ± S.E.) after hours of imbibition										
	0	1	2	3	4	5	6	7	8	24	48
N	0	0	0	1.0±0.7	7.0±3.3	21.0±3.1	37.5±3.9	48.0±3.4	64.0±4.0	93.5±3.9	95.5±2.0
LD	0	0	0	0	0	0.5±0.6	3.5±1.1	10.5±2.0	29.0±2.2	98.5±1.1	–

[a] From natural habitat in Jerusalem harvested 14. 7. 1967.
[b] Seed population matured under 16 h day length at temperatures of 27/22 °C harvested 11. 7. 1967.

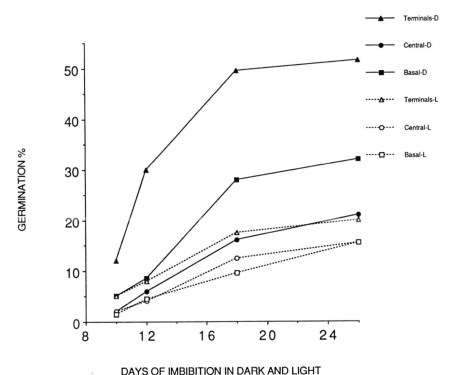

Fig. 121. *Mesembryanthemum nodiflorum* seed germination, after 10 to 26 days, of terminal (*T*), central (*I*) and basal (*B*) groups of seeds, in light (*L*) and dark (*D*). Seeds had been kept in capsules since harvested in summer 1972; tested in June 1991. (Gutterman 1990b)

not germinate during the first 4 years of afterripening. Each of the three groups of seeds differed in their germinability as well as in their light requirements for germination during the period of dry storage after ripening. After 6 years, however, seeds of the terminal group of a capsule achieved 99% germination in April under a combination of 35 °C during the day and 15 °C at night. They germinated to the same levels in both light and dark, at temperatures between 15 and 35 °C, after 6 – 8 years of storage (Gutterman 1980/81 a) (Tables 17, 18; Fig. 123). After 16 years, seed germination was promoted by continuous light during the time of imbibition. When germination was tested in October 1988, at 15 °C under continuous light, about 30% germination was observed. After 19 years of storage, the levels of germination were much lower in continuous light than in dark with short periods of illumination (Tables 17, 18; Figs. 121, 123). When germination was tested in June 1991 at 15 °C in darkness, after 19 years of storage, 54% of terminal seeds germinated, 21% of central and 32.5% of basal seeds. This is the highest percentage germination of central and basal groups ever observed. It is possible that seeds with long viability change their germinability according to time and season. Even under the optimal conditions, only a portion germinates (Table 54).

Table 54. *Mesembryanthemum nodiflorum* seed germination (% ± S.E.) after 35 days of imbibition at 15 °C in the light (L) and in the dark (D) after 19 years of dry storage of the seeds in their capsules. Harvested summer 1972, tested June 1991. (Gutterman 1990b)

Seed position in capsule	Light conditions during imbibition	Germination (% ± S.E.) after 35 days of imbibition
Terminals	L	22.0 ± 2.5
	D	54.0 ± 3.5
Central	L	16.5 ± 2.9
	D	21.5 ± 1.0
Basal	L	16.0 ± 2.5
	D	32.5 ± 6.0

An annual rhythm of high levels of germination in winter and very low levels in summer of the terminal seeds diminishes after ca. 16 years' storage (Gutterman 1980/81 b). Annual cycle changes in seed dormancy are also known in *Aphanes arvensis* (Roberts and Neilson 1982) and *Cyperus inflexus* (Baskin and Baskin 1978). *Asteriscus pygmaeus*, a desert annual (Sects. 2.2.4, 3.3.3.2), provides an interesting example of a long storage seed bank on the dead mother plant. The seeds are released from their capitulum by rain, whorl after whorl, over as many as 15 – 20 years. Only achenes disconnected from the capitulum can germinate. Achenes from the peripheral whorl are the first to be dispersed and the most central achenes are the last (Gutterman and Ginott, in press).

A question to be answered is: for how long are seeds in the seed bank still viable? What is the possible longevity of seeds? Seeds with long viability are important: (1) for the survival of the species under extreme and unpredictable conditions; (2) as an 'evolutionary memory' which increases the gene pool. Crossing of plants of different seed ages acts as a buffer to genetic changes in a population (Harper 1977; Silvertown 1982). The dormancy (Nikolaeva 1969) of seeds of many species enables them to remain in the seed bank for many years. Seeds of some species have been found to retain high viability for many years after maturation. They can be stored in the soil for as long as 90 years, as in the case of *Verbascum blattaria*, and still germinate to produce normal plants. Seeds of *Lupinus arcticus* have been found to be viable after 10000 years (Porsild et al. 1967) and of other species after hundreds of years (Mayer and Poljakoff-Mayber 1982; Fenner 1985). Low temperatures, dry conditions and the absence of oxygen, are important factors affecting viability during storage. Dry conditions (Went and Munz 1949) are the most important of the factors (Barton 1961; Harrington 1963, 1972; Styer et al. 1980; Kay et al. 1988). After 20 years, germination was still high (61%) in seeds stored under dry conditions. Under atmospheric conditions, however, there was a reduction of viability of similar seeds to 1% after 10 years (Went 1969).

It is possible that dry storage conditions with periods of wetting by dew may be important for repair mechanisms that increase viability. Long-term field experiments on the influence of natural storage conditions on the length of seed viability may be important for a better understanding of this stage in the life cycle of desert annuals. Such an experiment needs to focus on the influence of the position of the seed on the soil surface compared with storage at different soil depths.

Little information is available on seed viability after long periods of storage under natural desert conditions. However, the occasional mass appearance of species after many years may be the result of the accumulation of viable seeds over many years. Further study is needed for the better understanding of the strategies of germination and seedling survival of both desert annuals and perennials.

Where desert conditions are extreme, after one fall of rain which causes germination a long dry period is possible. Therefore, this may be an important survival mechanism when only a certain part of the seed population is dispersed (Sects. 3.3.3.2, 3.3.3.4) and germinates (Chap. 2; Sects. 5.1.2, 5.2.2). Together with long viability and the production of very large numbers of very small seeds, this is typical of many annual species of the Negev highlands.

4.3 Conclusions

As mentioned in Chapter 3, various plant species have different strategies for survival. Either they produce very small seeds in large numbers, or large seeds in relatively small numbers which are well protected from predation. Many species which produce very small seeds in large numbers also exhibit the opportunistic strategy of mass germination. Survival of annual species in the desert seems to be dependent upon the ability to ensure the existence of sufficient long-living seeds to ensure continuity.

4.3.1 Short-Storage Seeds

Different species may have completely different strategies of seed storage. There are species whose seeds mature at the time that is also suitable for germination. They germinate immediately after dispersal or after only a short period (16 days) of storage. Such seeds are recalcitrant and do not have the ability to dehydrate and survive (Berjak et al. 1989). An example is *Avicennia marina* (Verbenaceae), which inhabits parts of the seashore of the Sinai Peninsula (Evenari et al. 1985). Seeds of other species, such as the Mediterranean tree *Quercus calliprinos* (Fagaceae), mature in winter and do not have the ability to dehydrate. Therefore, they do not need additional water for germination. They can germinate immediately if they are dispersed to sheltered places where they do not lose too much of their water. They may be carried by rodents to their nests where they germinate if the embryo is not damaged. Seeds of the Mediterranean *Loranthus acaciae* (Galil 1938), as well as the East Sudanean *Viscum cruciatum* (Loranthaceae), inhabiting the Arava desert, become at-

166 Storage Conditions Affecting the Germinability of Seeds in the Seed Bank

tached to the beaks or legs of birds by the sticky fruit and are thus carried to branches of Acacia trees. They germinate after a short time as they do not require rain.

4.3.2 Long-Storage Seeds and Regulation of Germination

Seeds of some species mature in a single season but may remain in storage for one or more unsuitable seasons before germination and seedling establishment. Such 'hard seeds' exhibit different germination regulating mechanisms. These include the need for periods of high temperatures to overcome after-ripening, of fluctuations in temperatures during summer to 'soften' them (Quinlivan 1968). A gradual increase in humidity enables free water to penetrate through the hilum valve of the hard seeds (Hyde 1954). Scarification by floods of hard seed coats of seeds of species that inhabit wadis (Went 1953), an annual cycle of dormancy (Courtney 1968; Baskin and Baskin 1978; Roberts and Neilson 1982), as well as other mechanisms, ensure that germination will take place at the right time and in the right place. One of these is the thermo-inhibition of winter-germinating species. In these, high temperatures at the time of wetting prevents germination after unexpected summer rain. This mechanism may also prevent summer germination of winter-germinating plants in areas that receive both winter and summer rain (Gutterman 1990a).

4.3.3 Seeds with Strategies of Dispersal and Germination Over Many Years

1. *Dispersal by chance and age.* In *Blepharis* only up to 30% of capsules disperse immediately, even under optimal conditions and seeds are distributed by chance and age. The oldest are dispersed first (Gutterman et al. 1967). They may remain viable on the dead mother plant for many years, and some of them are dispersed periodically. The longer the time of storage, the shorter the time of wetting required before dispersal and the higher the percentage of capsules that are dispersed in one long rain event (Sect. 3.3.3.4).
2. *Seed dispersal according to the seed position in the capsule or capitulum.* (a) The seed bank of *Mesembryanthemum nodiflorum* remains in the soil after being dispersed during the rainy season following seed maturation. Each of three groups of seeds, according to their position in the capsule, is released after different amounts of wetting (Sects. 2.2.4, 3.3.3.2, 4.2.4). The terminal seeds are the first to be dispersed, and basal seeds the last. The terminal seeds germinate following about 4 years of afterripening in comparison with seeds from the other two groups which require many more years of afteripening but then germinate to higher percentages (Gutterman 1980/81a, 1990b). (b) The seeds of *Asteriscus pygmaeus* stored on the dead mother plant are released by rain, whorl after whorl, from the periphery to the centre, over many years (Sects. 2.2.4, 3.3.3.2, 4.2.4).

Conclusions

More research needs to be carried out on the short-term and long-term influences of storage conditions on seed viability and readiness for germination. It would be interesting to compare the influences of different locations on seed germinability, e.g. the location of the seeds on the dead plants, under or on the soil surface. Another question worth investigating is the viability of seeds located on the soil surface and the effect of wetting by dew ca. 190 times per year. The effect of high salinity and the presence of a repair mechanism should also be studied under natural conditions (Heydecker et al. 1973; Bewley and Black 1982; Liou 1987; Rao et al. 1987). These studies may lead to a better understanding of one of the most fascinating events typical of deserts: the mass germination and appearance of some species only once in many years (Gutterman 1983).

5 Environmental Factors During Seed Imbibition Affecting Germination

5.1 Introduction

5.1.1 Germination at the Right Time

The survival of annuals is dependent upon the germination of seeds at the right time of the season which gives the best chance for plant development and seed production. This increases the chance of continuity to a particular species in a particular habitat, especially for those annual species inhabiting extreme deserts. The more extreme the desert, the higher the percentage of annual species as well as the percentage of cover by annuals. This is especially pronounced in extreme deserts in seasons when rainfall is above the annual average of the area (Evenari et al. 1982).

The seeds are the stage of the life cycle of higher annual and perennial plants most resistant to extreme environmental conditions. Seeds can survive and retain their ability to germinate for many years (see Chap. 4) (Went 1969; Bewley and Black 1982; Evenari et al. 1982; Mayer and Poljakoff-Mayber 1982) (see Sect. 4.2.4). They can survive under very high temperatures for long periods (Kurtz 1958; Quinlivan 1966) as well as in high concentrations of salinity, such as accumulate near the soil surface in the desert during the hot, dry summer (Friedman and Orshan 1975; Gutterman and Agami 1987). This enables desert annuals to remain as seeds over long periods and germinate when conditions are suitable for development. The most important mechanisms for the survival of the desert annuals, especially species from extreme deserts, are those that regulate the timing of germination and of the change from the most resistant stage of the life cycle – the seed, to the most sensitive one – the seedling. The more extreme a desert, the more is survival of annual species dependent upon the time and number of seeds that germinate following a 'promising' rainfall. What is typical of extreme deserts is the completely unpredictable, very low annual rainfall as well as the unpredictable distribution of rain in the rainy season or seasons (Sect. 1.2). One question is: what makes the seed sensitive to the right conditions for germination and able to "predict" and germinate when conditions for the survival of the seedling and development of the new plant are most favourable?

One very simple mechanism which ensures the germination of seeds at the right time and place is found in trees and shrubs such as *Dalea spinosa, Olneya tesota* and *Cercidium aculeatum* of the Mojave and Colorado deserts of North

170 Environmental Factors During Seed Imbibition Affecting Germination

America. These species inhabit wadi beds and their seeds have very hard seed coats. Only after a heavy flood in the wadi grinds the hard seeds together with sand and gravel, can the embryo break through the scarified coat. Germination, therefore, takes place at the right time and in the right place when enough water is available for germination as well as for seedling establishment (Went 1953) (Sect. 4.2.1.2). There are many other regulating mechanisms in other desert species, which ensure germination at the right time and place. Examples are found in *Blepharis* spp. (Gutterman et al. 1967) (Sect. 3.3.3.4), *Asteriscus pygmaeus* (Sect. 3.3.3.2), *Mesembryanthemum nodiflorum* (Gutterman 1980/81a, 1990b) (Sect. 2.2.7) (see also Koller 1954, 1972; Evenari et al. 1982).

5.1.2 The Genotypic and Phenotypic Regulation for Seed Germination

1. Seed germination may be regulated by genotypic inheritance. This increases fitness to the natural habitat by the ability to germinate at the right time of the season and in the right place. Various plant species have their own regulation mechanisms such as: (a) the minimum amount of rain that triggers germination; (b) range of temperatures or thermoperiods for germination; (c) light or (d) dark requirements; (e) photoperiods of light and dark or dark and light; (f) chemical barriers such as germination inhibitors or (g) physical barriers such as hard seed coats; (h) embryo dormancy; (i) annual cycle of dormancy; (j) thermodormancy and thermo-inhibition; and (k) synaptospermic mechanisms which regulate seed dispersal in time and space, as well as (l) the time of germination. These germination regulation mechanisms may increase the chance of seedlings to survive by germinating during or following suitable rainfall (Evenari 1952, 1961, 1965b; Borthwick et al. 1954; Borthwick and Hendricks 1961; Barton 1965a, b; Lang 1965; Stokes 1965; Wareing 1965; Courtney 1968; Nikolaeva 1969; Burdett 1972; Koller 1972; Rollin 1972; Taylorson 1972; Hegarty 1973, 1978; Thompson 1973a, b; Toole 1973; Vidaver and Hsiao 1975; Duke et al. 1977; Heydecker 1977; Baskin and Baskin 1978; Bewley and Black 1978, 1982; Ungar 1978; Bewley 1980; Karssen 1980/81; Mayer and Poljakoff-Mayber 1982; Roberts and Neilson 1982; Cantliffe et al. 1984; Fenner 1985; Côme and Corbineau 1989; De Greef et al. 1989; Egley 1989; Hilhorst and Karssen 1989; Karssen et al. 1989; VanDerWoude 1989).
2. On the other hand, phenotypic influences may increase a diversity by mechanisms which ensure that, even under optimal conditions for germination, only a portion of the seed population of a particular species will germinate after any one rain event, or even during one season (Chaps. 2–4).

The percentage of the seeds in a 'seed bank' which are 'ready for germination' after a particular rainfall is a result of genotypic inheritance. It is also dependent upon phenotypic influences, such as the history of each of the seeds. It starts even from the position of the seed in the dispersal unit from which the mother plant of the particular seed originated (Table 16) (Datta et al. 1972b). Also involved are conditions of maturation, as well as phenotypic influences

Introduction 171

during the time of seed development (Chap. 2), dispersal mechanisms (Chap. 3); the storage (Chap. 4) and conditions of imbibition.

It might be expected that the better the fitness of a species to its habitat, the higher will be the 'predictability' of germinating at the right time of the right season. A more complex question arises: in the unpredictable environmental conditions of extreme deserts (Chap. 1), to what degree plant species are able to 'predict' the right time for germination. If the answer is in the affirmative for some species, certain environmental signal, or signals, must lead in the right direction by using 'low-risk' strategies. If the answer is negative for other species, there must be mechanisms that ensure survival in spite of the fact that they take a 'high risk' in predicting when the right time has arrived for germination.

5.1.3 High- or Low-Risk Strategies for Seed Dispersal and Germination

As can be seen from Chapters 2, 3 and 6, it would seem that: (1) in plant species with a 'cautious' strategy of dispersal and germination, predictability to germinate at the right time is high with a relatively 'low risk'; (2) plants with an 'opportunistic' strategy of dispersal and germination have a low predictability of germinating at the right time and a relatively low chance of seedling survival, plant development and seed maturation; (3) other species have a combination of low and high risk in their dispersal and germination mechanisms. If this is so, (a) there must be signals for plants with a 'cautious strategy' to germinate at the right time and (b) appropriate survival mechanisms in 'opportunistic' species (Fig. 122).

It has already been mentioned in Chapter 3 that some of the plant species inhabiting extreme deserts disperse their seeds after maturation without any limitations. One of these is *Schismus arabicus*, which produces a very large number of extremely small seeds (Sect. 3.1) and has a 'high risk' strategy of dispersal and germination. Its seeds germinate in a wide range of environmental conditions and even after relatively little rainfall (10 mm), thus giving the seedlings little chance of establishment and reproduction (Table 63). This compares with other species which need far more specific conditions for both seed dispersal and germination, or a 'cautious strategy' for both (Koller and Roth 1964; Evenari et al. 1977; Loria and Noy-Meir 1979/80; Gutterman 1980/81 a; Venable and Levin 1985; Venable et al. 1987; Beneke 1991).

'Safe' vs 'risky' options for germination are found in different plant species (Ellner and Shmida 1990). Species, such as *Blepharis* spp., which have more cautious strategies, regulate the dispersal of their seeds from the dead parts of the mother plant over many years (Sect. 3.3.3.4) (Gutterman et al. 1967) and *Asteriscus pygmaeus* (Sect. 3.3.3.2). Seeds of others are not dispersed at all and germinate in situ. These are exemplified by the subterranean propagules of *Emex spinosa* (Sect. 2.2.1) (Evenari et al. 1977) and the achenes of *Gymnarrhena micrantha* (Sect. 2.2.1). These amphicarpic plants have a dual strategy because their aerial 'seeds' are dispersed and much less 'cautious' than the

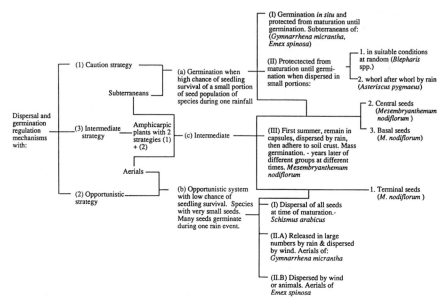

Fig. 122. Dispersal and germination strategies in some of the more conspicuous species inhabiting hot deserts such as the Negev Desert highlands

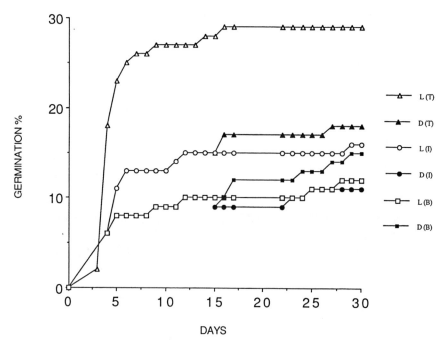

Fig. 123. *Mesembryanthemum nodiflorum:* seed germination of terminal (*T*), central (*I*) and basal (*B*) groups of seeds, in light (*L*) and dark (*D*) during 30 days. Seeds had been kept in their capsules since harvested in summer 1972. Tested in October 1988. (After Gutterman 1990a)

Introduction

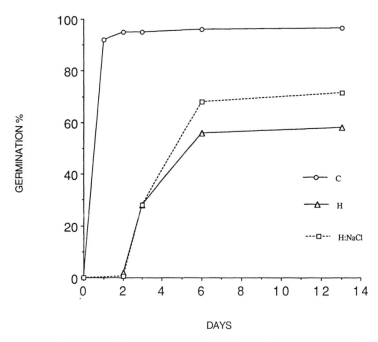

Fig. 124. Influence of high temperatures and NaCl on the germination of *Carrichtera annua* in continuous light. Seeds were germinated at 15 °C (*C*). Other seeds were exposed to 35 °C for 24 h and then transferred to 15 °C (*H*). Another group of seeds was imbibed for 24 h in 0.3 M NaCl during exposure to the high temperature (*H*:NaCl). (After Gutterman 1990a)

subterranean ones (Sects. 2.2.1, 3.3.3.2, 3.4) (Koller and Roth 1964; Gutterman et al. 1967; Evenari et al. 1977; Loria and Noy-Meir 1979/80). The three groups of seeds from the same capsule of *Mesembryanthemum nodiflorum* have been found to have different germinability, which changes during the years of storage (Sect. 2.2.7; Tables 17, 18; Figs. 121, 122, 123) (Gutterman 1980/81a, 1990b).

It would seem that the importance of the survival mechanisms of annual desert plants with the 'cautious' strategy is to ensure germination of only a portion of the seeds at any one time and only after a heavy fall of rain during the season which will give the seedling a chance to develop. This will enable the plant to finish its life cycle by producing enough seeds for survival in the future — a positive population balance.

One of the most fascinating questions is whether a plant species exists with a 'cautious' strategy of germination that prevents germination of the entire seed population unless environmental conditions are suitable for successful seedling development and subsequent seed reproduction. This includes the soil being moist enough for completion of the life cycle from seedling until seed maturation. What are the mechanisms of such a strategy? Do the same mechanisms exists in all these species or does each species, in its own way and through different mechanisms, arrive at the same result? The dilemma is that if the 'cautious strategy' is so important for survival why does it exist in so few species? Why do so many have 'opportunistic strategies' in which many seeds germinate after there has been a little amount of rainfall (10 mm), giving these seedlings a low chance of survival over a very short period of time? It would appear that the majority of such species have some combination of survival mechanisms which combine some elements from each of these two extreme strategies. Each species has its own proportional combination, but current knowledge on this subject is limited to very few species. There are some in which the regulation of germination mechanisms give the seedling the chance to survive only until the next rain and provided that the dry period between one big rainfall until the next is not too long. Such survival mechanisms are unique for each species and extremely complex because they are a result of the following:

1. Genotypic inheritance, which includes five basic processes of evolution that are involved in the genotypic inheritance of species. These are, 'mutation, genetic recombination, natural selection, chance fixation of genes and reproductive isolation', together with the 'rich but finite gene pools' (Stebbins 1977). They increase the fitness of the plant species to its habitat and enable it to regulate the germination to the right time of the right season and in the right place. Went (1953) came to the conclusion that the amount of water and the temperatures are such important factors in the regulation of the germination of desert plants 'that the others can be virtually neglected in a primary approximation'. This regulating mechanism may act as a 'rain gauge' and is dependent upon the amount of rain that dilutes the germination inhibitors.

   Afterripening and annual cyclic rhythms of seed dormancy and germination are also very important regulators of germination of desert plants (Courtney 1968; Baskin and Baskin 1978; Roberts and Neilson 1982). Other effects on embryo dormancy include the physical barriers of the seed coat (Hyde 1954; Quinlivan 1966; Egley 1989) (Chap. 4), the chemical barriers such as germination inhibitors (Sect. 5.2.2), and many others (Chaps. 2, 3). All these may regulate the germination of seeds at the right time of the right season of the year giving the seedlings the best chance for seed reproduction.

2. Phenotypic influences which increase the diversity of seed germination in time (Chaps. 2, 3, 4);

3. These, together with the unpredictable environmental factors found in desert habitats (Chap. 1), show how complicated it is for plants to survive in their desert habitats. These, and other environmental factors, affect the germination process, from the beginning of seed wetting during rain and during the time of imbibition.

5.2 Water

5.2.1 The Minimum Amount of Rain That Engenders Germination in Desert Plants

From studies in the Negev Desert highlands, a rain of more than 10–15 mm is usually enough for germination of several annual plant species, if this rain arrives during the month when the monthly evaporation rate from free water space is at its lowest in the year (Figs. 6, 7) as well as the temperatures (Fig. 5) (Evenari and Gutterman 1976). Loria and Noy-Meir (1979/80) found that near Sede Boker *Schismus arabicus* needs about 10 mm of rain while *Spergularia diandra* requires a little more. This may be compared with *Erodium bryoniifolium*, which will germinate only after more than 25 mm of rain, but nearly 100% of the plants produce seeds. From the results mentioned above, it seems that the various plant species inhabiting the Negev Desert differ in their strategies of germination regulation. They, therefore, need different amounts of precipitation for germination and also have different percentages of seedling survival.

There are large differences between the amounts of rain required for germination by different species inhabiting the same habitats. These can be correlated with the amounts that each species needs in order to finish its life cycle or to survive to the next rain event. The first strategy exists in some species and the second in others, or there may be a combination of the two. Seedlings of various species require different amounts of rain for their establishment and to finish their life cycle. It seems that what is even more important than the amounts of rain is the amount of water input to the soil exceeding a certain threshold (Beatley 1974; Gutterman 1980/81a, 1982a, 1988a, 1989a; Danin 1983). The germination, establishment and seed production of *Schismus arabicus* in habitats I, II, III, IV are summarized (Table 63). In small runnels (habitat I) the highest number of seedlings were observed in two seasons, 1973 and 1974. In drier habitats (II), less and in the driest (III) or (IV) only a few or no seedlings appeared. The threshold for germination of *S. arabicus* seems to be 10 mm or less in a micro-habitat where runoff water accumulates (Loria and Noy-Meir 1979/80).

Germination was found in porcupine furrows on the hills around Sede Boker from 12th–13th November, only after 7.1 mm of rain, above which runoff water accumulates. No seedlings were found at this time in the soil between the furrows (Gutterman 1982a). Porcupine diggings of the Negev highlands provide a unique ecological system for the germination and development of annual plants as well as for geophytes and hemicryptophytes (Gutterman 1982c, 1987, 1988a) because of the accumulation of runoff water (Sects. 3.2.2, 4.1.2, 6.3) (Gutterman 1989a; Gutterman et al. 1990; Shachak et al. 1991).

Seedling numbers are dependent upon the amount of rain and the amount of water penetrating the soil in certain habitats on the hill slopes near Avdat. The development of the seedlings and seed production is dependent upon the distribution of rain, which affects the soil moisture content during the growing season as well on inter- or intra-specific competition (Evenari and Gutterman 1976) (Sect. 6.1.2) (Tables 61, 65).

On a flat area between Jericho and the Dead Sea, *Mesembryanthemum nodiflorum* seedlings were observed during winter 1972/73, 16 days after the first rain, in depressions where runoff water had accumulated after three rains. On the flat area seedlings were not usually numerous until after 11 rain events. The seeds of this species germinate only after the salts from the soil surface where the seeds are situated have been washed deeper into the soil through leaching and washing by a number of rains. This is a mechanism of a 'rain gauge' system in which seed germination is inhibited until a minimum amount of water moistens the soil. *M. nodiflorum* inhabits areas where very large amounts of salts accumulate on the soil surface. Under laboratory conditions its seeds did not germinate after imbibing water containing low concentrates of NaCl (Gutterman 1980/81a). Even so, in this extreme area, the chance to flower and produce seeds in depressions is much better than between the depressions (Fig. 125).

At Avdat, in a flat natural area an experiment showed that it is possible as a result of artificial irrigation for certain species, such as *Schismus arabicus* and *Malva aegyptia*, to germinate during the hot and dry summer, but only after receiving water equivalent to 90–200 mm of rain (the optimum amount was found to be 150 mm) (Gutterman 1986b). All this irrigation was provided

Fig. 125. Developed *Mesembryanthemum nodiflorum* plants in depressions and undeveloped seedlings between the depressions

Water 177

during one day. A rain event of about 10–20 mm is enough in the winter to cause germination in these species, in the summer the amount required is some ten times more (Gutterman 1981).

In three species of annual plants that grow in the northern Chihuahuan desert of North America, seeds germinate after a rainfall of more than 10 mm. The greater the simulated rainfall, from about 15 to 45 mm, the higher the percentage of germination. Summer rain is abundant and reliable, while the winter and spring rain is less reliable. The summer annual *Pectis angustifolia* does not possess innate dormancy, while in the two winter annuals tested, *Lappula redowskii* and *Lepidium lasiocarpum*, germination under environmental conditions of increasing uncertainty resulted in an increase in innate dormancy (Freas and Kemp 1983). As only one summer and two winter species were observed, it is difficult to come to a general conclusion until more information is known about other species.

Juhren et al. (1956) showed that in the Joshua Tree National Monument desert of California 5.2–15 mm is the minimum amount of rain required for the germination of winter annuals during the autumn. If the first rainfall is large, many seeds germinate but, if the first rain is only 10 mm, good germination will take place only if there is another fall of rain of 5–6 mm a few days later.

Germination occurs in *Chaenactis fremontii* seeds when the temperature during the day is 12–30°C, and the temperature at night is 4–10°C, with rainfall that is always more than 10 mm, unless there has been previous rain that season of about 6 mm. *Chaenactis caryphoclinia* seeds never germinate at the required temperatures if the first rain is less than 10 mm although, on one occasion, germination appeared after 2 mm of rain preceded by snow.

Seeds of *Calyptridium monandium* germinate only after a first rain of 40 mm, whereas a late rain only needs to be of 11 mm. Seeds of several other species require less than 12 mm for the first rain and about 7 mm for the second rain. Germination of seeds of *Coreopsis bigelovii* occurs after 12 mm of rainfall and after a fourth rain of 6.6 mm. There is a large germination after rains of 30–43 mm. In *Amaranthus fimbriatus*, a summer annual, it was found that when the temperature during the day was between 28–35°C and at night between 12–20°C, an initial precipitation of 6 mm was sufficient for germination. A second summer rain, but of 23–50 mm, will also bring about germination. Seeds of *Boerhavia spicata* will not germinate in the first rain but do so after the second summer rain (see also Tevis 1958b, c). In the Australian deserts, seeds of winter-germinating Compositae require 15 to 20 mm of rain for germination and the time for germination to take place is 2–5 days. Seeds of summer-germinating grasses require at least 25 mm of rain and germination occurs after 24 h (Mott and Groves 1981).

The amount of rain required for seed germination during autumn and winter for winter annuals, and during summer for summer annuals in deserts which have summer and winter rains, is almost in the same range in the majority of species observed. Temperature is also important as a regulating mechanism and the time needed for imbibition also varies (Juhren et al. 1956). From the foregoing it can be seen that the minimum amount of a single rain event that triggers germination is nearly always the same (10–15 mm) for species with the 'opportunistic' and high risk strategy of germination in the hot deserts of Australia, South Western North America, and in the Saharo-Arabian region of the Negev Desert (in winter but not after artificial summer irrigation). The other species with the more 'cautious' and lower risk strategy germinate after more than one fall of rain and a heavy precipitation of 25 to 50 mm. These observations can be summarized as follows:

1. Various plant species, even those inhabiting the same habitat, require different minimum amounts of rain for germination. There are seeds of species that germinate after 10 mm or even less rain in the different hot deserts of the world and others require at least 25 mm, or even 50 mm (Freas and Kemp 1983; Juhren et al. 1956; Mott and Groves 1981). This is connected with different strategies of survival, such as 'opportunistic' and high risk strategies, for species that germinate following ca. 10 mm of one rainfall, and 'caution' and lower risk strategies for germination and seedling development for others that germinate only after much more rain.

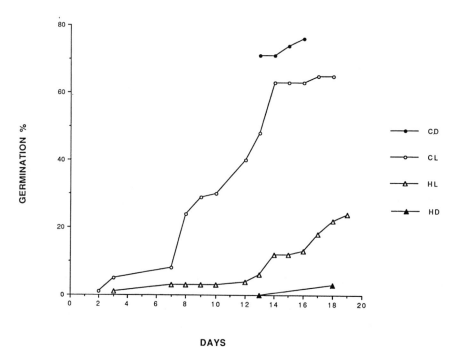

Fig. 126. Influence of high temperatures and light on the germination of *Cheiridopsis* species No. 55 which germinated at 25 °C in the dark (*CD*) and in the light (*CL*). Other groups of seeds were kept at 45 °C for the first 24 h and were then transferred to 25 °C, in continuous light (*HL*) and in continuous dark or in the dark with short periods of illuminations after the first counting (*HD*). ANOVA test; p = 0.0001 germination after 18 days: (*CL* vs *HL*)*; (*CL* vs *HD*)*; (*CD* vs *HD*)*; (*HL* vs *HD*)**; * = significant at 95% by Scheffe F-test; ** = significant at 95% by Fisher PLSD (Gutterman 1990a)

2. In a desert such as the Negev where rain only falls in winter, some annual species will germinate even during the hot and dry summer. They require a much higher minimum amount of irrigation (×10) in comparison with the minimum amount of rain needed during winter for this species. But other plant species will not germinate in spite of the large amounts of water supply (Gutterman 1981).
3. In deserts with both winter and summer rains the amounts required for germination are similar, but temperature is a more important regulator for the germination of winter and summer annuals (Juhren et al. 1956). As will be shown below, there are different germination mechanisms in plant species originating from deserts receiving either winter or summer rain, as seen in species from the Karoo desert. Species inhabiting areas receiving winter rains may have seeds which are thermo-inhibited when the first 24 h of imbibition is at high temperature. In other plants which inhabit areas receiving summer rains, high temperatures during the first 24 h of seed imbibition accelerates germination and increases the levels of the percentage of germination (Sects. 5.4.3, 5.4.4; Figs. 126, 127) (Gutterman 1990a).
4. The amount of rain is a factor easy to measure. But, depending on the type of soil, the penetrability of the soil, slope depth, direction, rocks, soil crust, stone-soil cover, runoff water and depressions, differing amounts of water accumulate in different micro-habitats during and after a rain event. These

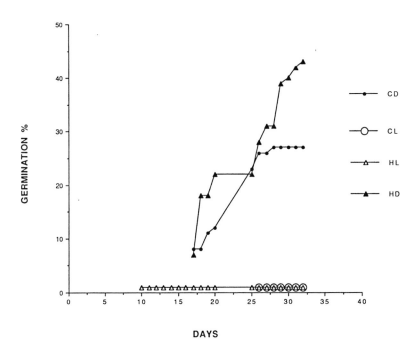

Fig. 127. Influence of high temperatures and light on the germination of *Bergeranthus scapiger*. ANOVA test; p = 0.0001 germination after 32 days: *(CD* vs *HD)**; *(HL* vs *HD)**; *(CD* vs *CL)**. For further details see Fig. 126 (Gutterman 1990a)

180 Environmental Factors During Seed Imbibition Affecting Germination

differences in distribution and penetration of water have a considerable influence on the germination of seeds depending on their position in different habitats. In depressions where runoff water accumulates, the dilution of salinity and germination inhibitors is faster, and seeds in these places are able to germinate after a relatively light fall of rain. This is not the case for seeds that are situated on the soil crust where runoff water does not accumulate. The soil near the surface of depressions remains moist for longer periods, especially where more plant remains accumulate. This enables seeds which require longer periods of imbibition to germinate (Gutterman 1990a). (Sect. 1.2.4; Fig. 11) (Evenari et al. 1982).

5. Some species will not germinate after a single fall of rain, but will germinate after the second or the following rains. Up to 11 rain events may be necessary for the germination of *M. nodiflorum*, which requires leaching of the salts from the soil surface. This acts as a 'rain gauge' which regulates germination to the proper time. Germination takes place much earlier in depressions where runoff water accumulates (Gutterman 1980/81a). The leaching of an inhibitor is also necessary for the germination of other species (Koller 1954; Datta et al. 1970; Evenari et al. 1977). (See Chap. 2).

5.2.2 Germination Inhibitors as 'Rain Gauges'

5.2.2.1 Germination Inhibitors in the Seeds or Dispersal Units

Dispersal units, such as the aerial propagules of *Emex spinosa*, contain water-soluble germination inhibitors. After imbibition for 7 days at 20 °C, 7% of untreated propagules germinated in light, and 24% in dark. Propagules which had been leached for 48 h in distilled water germinated to much higher levels: in light (36%) and in dark (60%) after the same time of imbibition (Table 7). Propagules leached and dried, then allowed to imbibe water for 7 days at 20 °C, germinated in light (36%) and in dark (58%) but, in leachate, only 7% germinated in light and 28% in dark (Evenari et al. 1977).

Dispersal units of *Aegilops geniculata* (= *A. ovata*) also contain germination inhibitors such as Monoepoxyligananolide (MEL) which dissolve in water (Lavie et al. 1974; Gutterman et al. 1980). The germination of achenes of *Lactuca sativa* in leachate from hulls of spikelet (a) was 48%, of (b) 36% and of (c) 29%. Of the control which imbibed water 94% germinated at 26 °C in the light (Fig. 18; Table 13; Sect. 2.2.3) (Datta et al. 1970). Germination inhibitors such as coumarin and phenolic substances were found in the dispersal units of *Trigonella arabica*, which are unopened pods (Koller 1954; Lerner et al. 1959; Mayer and Poljakoff-Mayber 1982) (Table 55). Inhibitors have also been found in the fruit wings of *Zygophyllum dumosum* (Table 56) and in the dispersal units of *Rumex cyprius* and *Atriplex halimus* (Table 57). The germination of *Atriplex dimorphostegia* seeds from both types of dispersal units is inhibited by a water-soluble germination inhibitor that is diffused from the fruit bracts. This is possibly Cl which composes 42–44% of the husk of the dispersal unit

Water 181

Table 55. Germination (%) of scarified seeds of *Trigonella arabica* after imbibing water with or without empty pods, in the dark at 20°C (After Koller 1954)

Treatment	Germination (%)
With Pods	10
Water	93

Table 56. Influence of 2.5 g/100 cc leachate or a dilution (×2) of *Zygophyllum dumosum* winged fruit on the germination of Grand Rapids lettuce achenes after 16 h at 26°C in light or dark at 26°C. (After Koller 1954)

Treatment	Germination (%)	
	Light	Dark
Leachate 100%	0	0
Leachate 50%	71	24
Water	95	64

Table 57. Dispersal units of *Atriplex halimus* harvested in 1950 and 1952, washed or washed and water imbibed. The germination (%) (4 × 50) from separated fruit or from the dispersal units after 1–8 days of imbibition on 2–10 November 1953, at 20°C in dark. (After Koller 1954)

Harvest	Imbibition (days)	Germination in (%)			
		Naked fruit ('seeds')	Dispersal units		
			Control	Washed	Washed and imbibed
1950	1	10.5	0.0	0.0	4.5
	2	59.5	3.0	3.5	42.0
	3	68.0	19.5	30.0	57.0
	8	71.0	56.0	64.5	71.0
1952	1	21.0	0.0	0.0	4.5
	2	71.5	2.0	1.0	36.0
	3	82.0	28.5	37.5	62.0
	8	83.0	58.0	68.5	83.5

(Koller 1954, 1957). Germination of *Salsola inermis* does not begin unless the germination inhibitors have been washed from the winged dispersal unit. This ensures that fast germination in this species will only begin after the soil has received a minimum amount of water (Evenari et al. 1982). These germination inhibitors, therefore, act as 'rain gauges'. The question is whether the inhibitors regulate germination only during the first rains of the season following maturation or, as Went (1953) proposed, that seeds germinate when the leaching of the inhibitor or inhibitors is faster than its accumulation. Went also proposed that diffusible growth-promoting substances are leached out of seeds at a slower rate than the inhibitors are increased. Until now, there has not been sufficient evidence to prove or disprove this interesting hypothesis (Mayer and Poljakoff-Mayber 1982).

In the Negev some experiments with artificial irrigation of natural soils have shown a decrease in seedling numbers in treatments with more than the annual average amount of rain (Gutterman et al. 1982).

5.2.2.2 Germination Inhibitors in the Soil Act as 'Rain Gauges'

Plant chemicals as inhibitors in plant litter influencing other plants (Allelopathy). Plant litter of trees and shrubs which accumulates beneath the plants has been found to inhibit the germination of seeds (Allelopathy) (Molisch 1937). In this way, competition is prevented. One of the reasons for mass germination after a forest fire is the disappearance of plant litter in which water-soluble germination inhibitors accumulate (Evenari 1961; Naveh 1973, 1974; Dick-Peddie and Alberico 1977; Keeley et al. 1985, 1989; Gill et al. 1986; Keeley 1986, 1987). After a fire in the Chaparral vegetation of California charred remains even stimulate the germination of *Emmenanthe penduliflora* (Wicklow 1977) as well as of *Antirrhinum coulterianum, Chaenactis artemisiaefolia* and *Eriophyllum confertiflorum* (Keeley et al. 1985).

Germination of seeds of *Erica hebecalyx* Beth. collected in the George area (Cape Province, South Africa) was stimulated by factors associated with fire such as light treatments at 80 °C for 3 min or even 96.5 °C for 3 min, as well as by 1 % ethylene for 24 h or 20 ppm ammonia on imbibed or dry seeds. These factors have no influence on seeds of related species *E. sessiliflora* L. f. which were collected from the same area (Van der Venter and Esterhuizen 1988).

In extreme deserts, most of the soil surface is completely bare before the rainy season. In this case, the high salinity which accumulates above or near the soil surface during the summer has an important ecological effect.

High salinity on and/or near the soil surface as a germination regulator. Evaporation from the soil surface causes an accumulation of salts near and above the surface of desert soils during summer. The dilution of salts from the layer in which seeds germinate acts as a 'rain gauge' which prevents germination of seeds that are sensitive to high salinity. The seeds of *Mesembryanthemum nodiflorum*, which inhabits soils of high salinity, only germinate after several rain events have washed the salts from the topsoil. This ensures that germination occurs after the soil has reached a degree of moisture to a minimum depth. This helps the establishment of the seedlings. From the field observations already mentioned, it was found that germination occurs only after about 11 days of rain. Seeds only germinate when the salinity of the soil has been reduced by infiltration to a certain depth so that the soil surface has a low saline content (Sects. 2.2.7, 5.2.1) (Gutterman 1980/81a). In this case the same regulating system acts on *M. nodiflorum* seeds at the beginning of each rainy season.

Helianthemum vesicarium Boiss., a species of Irano-Turanian origin, and *H. ventosum* Boiss. (Cristaceae), a Saharo-Arabian plant, are perennial shrubs which inhabit the hills around Sede Boker. The former grows on north-facing and the latter on south-facing slopes. Soil moisture is higher on north-facing slopes, and the temperature and soil salinity are lower than on south-facing slopes. Conductivity of the soil crust, 36 h after the 14 mm of rain, was 167.0 µmho/cm on a north-facing slope and 210.2 µmho/cm on a south-facing slope. After 196 h, conductivities were 223.5 and 285.2 µmho/cm on the north-facing and south-facing slopes, respectively. Chloride measurements taken from the same soil samples showed large differences between the two slopes: 7.5 ppm on the north-facing slope and 41.7 ppm on the south-facing slope (Gutterman and Agami 1987). Friedman and Orshan (1975) also found higher concentrations of Na^+ and Cl^- on south-facing slopes near Sede Boker. Scarified seeds of *H. ventosum* germinate faster and at higher temperature and salinity than do those of *H. vesicarium*. These adapt the two species to their respective habitats. The germination percentage of unscarified seeds of both species is very low (Figs. 128–130).

There were substantial differences between the two species when seeds imbibed water containing different concentrations of NaCl. Scarified seeds of *H. vesicarium* germinated best (68%) in distilled water, but only 20% in 42.8 µmho/cm (0.25%) NaCl, while no seeds germinated in NaCl concentrations above 85.5 µmho/cm (0.50%). In contrast, only 52% of scarified seeds of *H. ventosum* germinated in distilled water. Germination was best in 42.8 µmho/cm (0.25% NaCl) (about 68%). In 171.0 µmho/cm (1% NaCl), 38% of seeds germinated. Only above 427.5 µmho/cm (2.5% NaCl) was there no germination at all. The large differences in germination between the two species with regard to salinity and temperature are an adaptation to their respective habitats (Figs. 128, 130) (Gutterman and Agami 1987).

Seeds of *H. vesicarium* and *H. ventosum* collected from two elevations in the Negev highlands were tested for germination at constant temperatures from 5–40°C and at various levels of salinity at 15°C. One of the habitats was in the transition zone between the Irano-Turanian and the Saharo-Arabian regions at an altitude of 470 m near Sede Boker. The other habitat was in the Irano-Turanian region at 920 m above sea-level, near Mitzpe Ramon, in less extreme desert conditions. Here the two species occur on the same slope in the same plant community. The distance between the two habitats is 40 km.

Differences in germination of scarified seeds of the two species from both habitats were found to exist at constant temperatures, as well as in different solutions of NaCl. There are clear differences in germinability between seeds from the two localities and between the two species. Seeds of *H. ventosum* from populations at 470 m invariably germinated faster than seeds from popu-

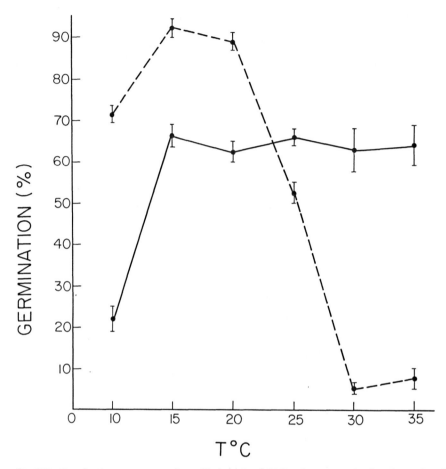

Fig. 128. Germination percentage of scarified seeds of *Helianthemum vesicarium* (– – –) and *H. ventosum* (———) after 6 days in continuous light at constant temperatures in the range 10–35 °C: mean ± S.E. (4 × 50 seeds). (Gutterman and Agami 1987)

lations at 920 m. The same results were observed among the populations of *H. vesicarium*, but their germination was slower. The populations of *H. ventosum* from more extreme desert conditions at 470 m germinated more quickly, while *H. vesicarium* from 920 m germinated slowest. The two populations of *H. vesicarium* also differ in the colour of their flowers and in the properties of their seed germination. The flower colours are the same in *H. ventosum*, but seed germination differs (Fig. 129) (Gutterman and Edine 1988).

After a period of 72 h in concentrations of NaCl from 0.25% (42.8 µmho/cm) to 1.75% (299.2 µmho/cm), there was a difference in germination among the four groups. *H. ventosum* from 470 m germinated best in distilled water and in salt concentrations from 0.25% to 1.75%, while *H. vesi-*

Water

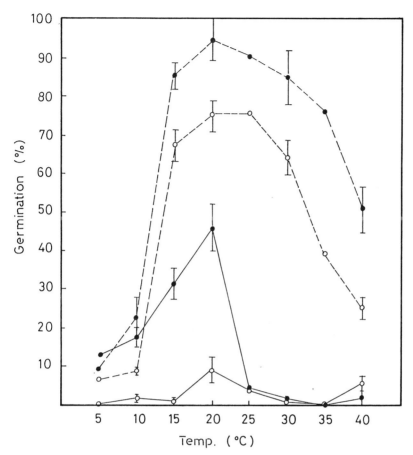

Fig. 129. Cumulative germination of scarified seeds after 72 h in continuous light and at constant temperatures of 5, 10, 15, 20, 25, 30, 35 and 40 °C; – – – *Helianthemum ventosum;* ——— *H. vesicarium;* ● 470 m; ○ 920 m. (Gutterman and Edine 1988)

carium from 920 m showed the poorest germination. Differences were found between the populations from the two elevations along the salinity gradient from distilled water (control) to a concentration of 1.75% NaCl with one exception. There was a large difference between the two populations of *H. vesicarium* in the control treatment, although this differerence diminished as the concentration of salt increased (Gutterman and Edine 1988).

As mentioned, seeds of both species of *Helianthemum* from 470 m germinated faster than seeds from populations at 920 m. As germination is faster, root penetration of the seedlings takes place earlier and the chances of seedling establishment is higher, especially in more extreme desert conditions where the upper layer of the soil dries up rapidly after a fall of rain.

Carrichtera annua seeds were allowed to imbibe water for 24 h in 0.3 M NaCl at high temperature (H). They were then washed and transferred to 15 °C

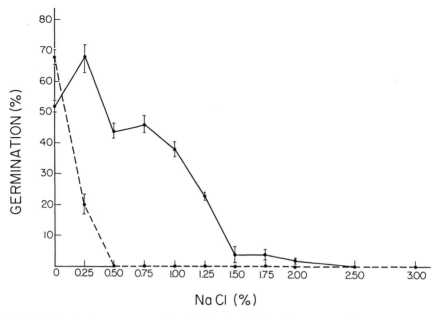

Fig. 130. Germination percentage of scarified seeds of *Helianthemum vesicarium* (– – –) and *H. ventosum* (———) after 6 days in continuous light at 20 °C as a function of NaCl concentration: mean ± S.E. (4×50 seeds). (Gutterman and Agami 1987)

in distilled water in light (L). The inhibitory effect was then less intense, and after 13 days the percentage of germination was 72% [HL (NaCl)] in comparison with 58% (HL in the absence of NaCl) and 97% in the control, (CL) germinated at 15 °C in constant light. Seeds of *C. annua* are mucilaginous and adhere to the soil surface after they are dispersed by rain (Figs. 124, 131, 132). These seeds are found in places where high salinity accumulates in summer. It is interesting to note that they germinate to a high percentage even after a rain under high temperatures if within 1–2 days a heavy rain washes the salinity from the soil surface (Gutterman 1990a). Beneath the canopy of *Tamarix* trees there is a highly saline accumulation from the salts from the secretory cells of the leaves (Fahn 1967). This prevents the germination of other plants in such locations and also prevents competition.

5.3 The Period of Wetting for Germination

One of the genotypic inheritances found in some desert species is concerned with the different length of time of water-imbibition required by their seeds before germination. The seeds of some desert species germinate slowly and possess specific light requirements. In addition, germination is dependent, to some extent on the amount, but also on the distribution of rain. This must

Fig. 131. *Carrichtera annua.* Mucilaginous seed adhering to soil surface by mucilaginous connections; ×52

Fig. 132. *Carrichtera annua.* High magnification to show mucilaginous connections to soil particles; ×104

moisten the soil surface continuously for as long as is required for seeds to germinate. The seeds of *Artemisia sieberi* require light and more than 16 days of imbibition for germination to take place. This, of course, is a very rare occurrence in extreme deserts (Evenari and Gutterman 1976) (Sect. 1.2.1.2).

Some perennial Aizoaceae of South Africa can also be divided into two groups according to whether they have fast or slow germinating seeds (Gutterman 1990a). In contrast, there are species whose seeds germinate after less than 1 h of imbibition. Examples of these are *Blepharis* spp. (Gutterman et al. 1967) and *Salsola kali* (Wallace et al. 1968).

5.3.1 Fast-Germinating Seeds

Seeds that germinate after a very short period of imbibition have been found in *Blepharis* species (Sects. 3.3.3.4, 4.3.3). In these, the radicle penetrates the soil within a few hours. Under optimal conditions, the radicle begins to appear with a geotropical bend within even less than 1 h. This fast germination enables the seeds of *Blepharis* to germinate during the same rain event that disperses the seeds, and to attain a radicle of about 50 mm in length within 24 h from the time of dispersal. Delay of seed dispersal (Sect. 3.3.3.4) until after a large amount of rain, together with rapid germination at a range of temperatures from 8 to 40°C, in light and dark, are a very important survival mechanism. It results in a high percentage of survival in the seedlings of this species (Table 45) (Gutterman et al. 1967; Gutterman 1972). Germination is even more rapid in *Salsola kali*. After only 29 min of imbibition, some of the seeds begin to germinate (Wallace et al. 1968). A connection between dispersal and germination similar to that found in *Blepharis* spp. has also been found in *Anastatica hierochuntica*. Some seeds are dispersed by rain after about 1.5 h of wetting and germinate within about 6 h of imbibition (Friedman et al. 1978; Gutterman 1982a).

5.3.2 Slow-Germinating Seeds

The seeds of some desert plants need a long period of wetting and light before they can germinate. This is the case with *Artemisia sieberi*, a perennial shrub typical to loess soil, found in the Irano-Turanean plant geographical region. Germination does not take place until after about 16 days of imbibition in light even under optimal conditions for germination in the laboratory. The mucilaginous seeds adhere to the soil surface shortly after dispersal in December and January when, in most years, the majority of the rain in the Negev highlands falls (Fig. 139) (Sect. 1.2.1.2).

In the winter of 1963/64, conditions of constant wetting of the soil surface prevailed for such a long period of time that mass germination of *A. sieberi* occurred in all parts of the Negev Desert highlands. An average 43/m² seedlings were counted in Avdat Evenari Research Station on a stony Hammadoid north-facing slope. In one m^2 plot a maximum of 159 seedlings appeared, but in another only four. In spite of this large diversity, such mass germination has not been observed on any other occasion during the last 30 years, possibly because such a long period of imbibition on the soil surface is rare. Only once in several years does the mass germination of *A. sieberi*, and other perennial species, enable the gaps to be filled where old shrubs have disappeared over the years. After the mass germination described, the vegetation took 9 years to reach full recovery and return to its original distribution and size (Evenari and Gutterman 1976). The mass appearance and survival of *Zygophyllum dumosum* has also been analyzed in 'age groups' (Evenari et al. 1982).

Another example is *Artemisia monosperma* (Koller et al. 1964; Koller 1969). In this species, the longer the period of wetting, the higher the percentage of germination. This perennial is typical of active sand dunes and stable sand fields in the Saharo-Arabian and Irano-Turanian plant geographical region. After 10 days imbibition the percentage of germination is high. Germination takes place below a minimum depth under the sand surface, where the humidity is suitable and light intensity is still favourable. In this type of habitat there are three limitations to germination. The light intensity decreases with depth, wavelength increases with depth (Sect. 4.1.3.2), and water content increases with the depth.

5.3.3 Air, Mucilaginous Seeds (Myxospermy) and Seed Germination

Seeds of some desert species develop a mucilaginous layer when wetted. At least in *Blepharis* spp., this layer acts as an O_2 barrier which prevents germination until the excess water has disappeared. This mechanism enables the seeds of plants inhabiting wadi beds to be dispersed by floods without being damaged (see Tables 46–48, 58a, b; Figs. 110, 111; Sects. 3.3.2.5, 3.3.3.4) (Gutterman et al. 1967; Witztum et al. 1969).

Germination of the mucilaginous seeds of other species of the deserts of Israel is not known to be inhibited by excess water. These seeds adhere to the soil surface after being dispersed and the mucilaginous layer, when dry, prevents seed eaters from collecting the seeds (Figs. 106–109). When wet this mucilaginous layer acts as a 'seed-bed' connecting the seed to the wet substrate (Sects. 3.5, 3.7).

5.4 Temperatures

Some species have seeds that germinate in a wide range of temperatures (8–40 °C). This has been found in *Blepharis* spp. (Gutterman 1972). Other seeds germinate only in a narrow range (Sect. 5.4.1). Imbibition at low temperatures (3–15 °C) appears to cause injury in some species. This includes changes in membranes. It is afterwards expressed in reduced growth and development (Bewley and Black 1978).

190 Environmental Factors During Seed Imbibition Affecting Germination

Table 58a. Effect of percentage of oxygen in the air above the seeds on the germination and cotyledon colour of seed under 3 ml of water at 25 °C. (Witztum et al. 1969)

Oxygen (%)	Germination (%)	Cotyledon colour	Range of root length (mm)
20	10	Yellowish	0 – 5
40	60	Yellow to yellow-green	0 – 29
60	100	Light-green	8 – 32
80	100	Green	8 – 27
100	100	Green	6 – 32

Table 58b. Germination (%) at 25 °C under 3 ml of water at varying percentages of oxygen in the air above the seeds. (Witztum et al. 1969)

Oxygen (%)	Whole seeds	Seeds minus mucilage	Seeds minus entire integument
2	0	0	0
5	0	0	20
10	0	20	100
20	10	80	100

Protein denaturation, autocatalytic enzyme reaction, and changes in membranes affect germination rates at high and low temperatures. Various species, from the same area but inhabiting different slope directions and elevations, have been found to have different rates of germination at different temperatures. Examples are afforded by *Helianthemum vesicarium* and *H. ventosum* (Sect. 5.2.2.2; Figs. 128, 129) (Gutterman and Agami 1987; Gutterman and Edine 1988). The interaction between temperature and light resulting in photo-inhibition appears at temperatures of 21 – 27 °C in *Nemophila insignis*, but not at lower temperatures (Bewley and Black 1982). There are also high temperature effects on embryo dormancy, including thermodormancy and thermo-inhibition (Evenari 1952; Burdett 1972; Gutterman et al. 1972; Toole 1973; Cantliffe et al. 1984; Gutterman 1990a; Small and Gutterman 1991, 1992a). Low temperatures reduce the speed of germination and development of seedlings. They delay the germination of annuals and perennials in the Negev (Evenari et al. 1982), as in the Sonoran Desert (Tevis 1958b).

5.4.1 Temperature and Germination of Winter or Summer Annuals

In deserts receiving both winter and summer rains the temperature of the soil when the rain begins is a very important environmental factor for regulating seed germination. Some species germinate in winter when temperatures are low, while others germinate in summer when temperatures are high. Topsoil was collected from four natural desert areas in the Colorado desert and spread thinly over washed sand in flats. When subjected to artificial rain and to a variety of temperatures, many seeds germinated. With 27 °C day and 26 °C night temperatures, a typical summer vegetation of annuals developed, comprising *Bouteloua barbata, B. aristidoides, Pectis papposa, Amaranthus fimbriatus, Mollugo cerviana* and *Euphorbia micromera*. None of the typical winter annuals appeared. The latter appeared in the flats kept during the day at 18 °C and during the nights at 13 °C or 8 °C. None of the summer annuals grew there, but winter annuals like *Gilia aurea, Plantago spinulosa, Eriophyllum wallacei, Necacladus longiflorus* and *Bromus rubens* developed well. Another group germinates at intermediate temperatures. This includes plants that germinate under natural conditions in the late autumn or at the beginning of winter (Went 1949; Juhren et al. 1956).

5.4.2 Temperatures, Day Length and Seasonal Genoecotypes

Seasonal genoecotypes such as *Chenopodium album* grow at Chandigarh, India. The seeds of summer genoecotypes germinate and develop from March to September when high temperatures average 27 – 32 °C and day length 12.5 – 14 h. Those of winter genoecotypes germinate and develop and grow from November to April at an average temperature of 11 – 20 °C and day length of 10.5 – 11.5 h. The summer populations are diploid (2n = 18) and the winter populations hexaploid (2n = 54) (Ramakrishnan and Kapoor 1974).

5.4.3 Thermodormancy and Winter-Germinating Species

The southern area of the Sahara receives summer rains, while the northern area receives winter rains (Trewartha 1961; Griffiths 1972; Takahashi and Arakawa 1981). The Karoo Desert has an analogous climatic division in which the western part, parallel to the western coast of South Africa, receives winter rains when temperatures are low, while the central and eastern part receives mainly spring and autumn rains when temperatures are relatively high. Some areas, however, receive rain during both the cold and the hot season (Rutherford and Westfall 1986). Consequently, there are both winter- and summer-germinating species, and the following question arises: are there special mechanisms of seed germination in the winter-germinating plants which prevent them from germinating during the summer, in spite of the fact that in these areas there is usually a transient drop of temperature during the days of rain which follows the high temperatures at the beginning of a summer storm? High temperatures at the beginning of imbibition by the seeds of some species have an influence on the proportion of seeds which enter a stage of thermoinhibition or thermodormancy.

Studies on lettuce "seeds" (Borthwick et al. 1952, 1954; Evenari 1952; Gutterman et al. 1972; Saini et al. 1986; Small and Gutterman 1991, 1992a) as well as on *Lactuca serriola* L. "seeds" collected near Sede Boker (Small and Gutterman 1992b), show that exposure to supra-optimal temperatures during the first hours of imbibition induced thermodormancy. Even if then transferred to optimal temperatures, germination was inhibited. Such a mechanism exists in the seeds of winter-germinating species inhabiting desert areas that receive mainly winter or both winter and summer rains. Such mechanisms could prevent the germination after rains occurring in summer, even if temperatures drop temporarily after the rain.

Some perennial shrub species of Aizoaceae (Mesembryanthemaceae) from the Karoo were introduced as seeds and are growing in the Sede Boker 'Introduction Garden'. Seeds matured on these shrubs were tested for germinability. Those collected from plant species which originated in areas receiving winter rain were compared with those receiving summer rain. In *Cheiridopsis* spp., inhabiting areas of winter rain, imbibition of seeds at 45 °C for 24 h inhibited germination in the light and prevented it completely in the dark at 25 °C. Seeds of the perennial shrub *Cheiridopsis* spp. (No. 55, an unidentified species), which had imbibed water at lower temperatures (25 °C), germinated to 48% within 13 days in continuous light (CL) and to 71% in continuous dark (CD). In this species, there is a very strong germination inhibition. This was revealed in the seeds exposed to high temperatures (45 °C) during the first 24 h. Only 6% of the seeds germinated after 13 days of imbibition in continuous light (HL). After 18 days, percentage germination was only 3% in continuous dark (HD) and 22% in continuous light (CL). At 25 °C, however, 66% of the seeds germinated after 18 days under continuous light (CL) (Fig. 126).

The inhibitory effect of the first 24 h of high temperature is much more pronounced in the seeds of *Cheiridopsis*, than the thermo-inhibition effect found in cultivated lettuce 'seeds', or in *Lactuca serriola*. Under continuous light, the latter overcomes the inhibitory effect of high temperatures after 48 h. Seeds germinated to almost 100% within 24 h at the optimal temperature (Small and Gutterman 1992b). When seeds of *Lactuca serriola* collected in the desert near Sede Boker were exposed to 33 °C for 36 h, they did not germinate. But 18–22% germinated within the first 24 h when they were transferred to 20 °C in light. When imbibition of 0.1 M NaCl for 36 h at 33 °C occurred, there was no germination but, after transfer to lower temperatures and washing, about 90% germination was reached within 24 h under continuous light. There was no germination in continuous darkness.

These phenomena may play an important role in germination under desert conditions. Exposure to high temperatures prevents germination in saline soils. If the salt is diluted by large quantities of water, however, seeds will germinate even after being exposed, in light, to high temperatures.

5.4.4 Summer-Germinating Species

The opposite has been found in seeds of species which germinate after rainfall in areas receiving summer rains. High temperatures will even accelerate germination. In *Bergeranthus scapiger*, which inhabits areas of the Karoo desert receiving summer rain, differences in germinability were found after exposure of the seeds to high temperatures during the first 24 h of imbibition. Seed imbibition at 45 °C for 24 h did not inhibit but accelerated germination when the seeds were later transferred to 25 °C. All the species of the genus *Bergeranthus* inhabit areas receiving spring/summer and mainly autumn rains (Herre 1971). Almost no seed germination was observed in the light (0–1%) after 17 days, while only 7–8% germinated in the dark. Better germination was observed in the dark: after 32 days at 25 °C – (CD) the germination percentage was 27% and after exposure to high temperature during the first 24 h of imbibition (HD) it even increased to 43% (Fig. 127).

Fouquieria splendens which germinates after summer showers in the Joshua Tree National Monument, California, requires alternating temperatures of 40 °C/20 °C to reach the highest germination percentage. The daily exposure to 40 °C needs to be 12–16 h (Went 1948) (Sect. 5.4.7).

5.4.5 Different Strategies of Two Negev Plants

Seeds of *Carrichtera annua* (Cruciferae), and other annuals of the Negev highlands, are inhibited from germination more intensely by darkness than by light after imbibition at 35 °C for 24 h (Figs. 124, 131). The opposite is true in germination of *Reboudia pinnata* (Cruciferae) seeds which are also mucilaginous, are dispersed by rain and adhere to the soil surface by their mucilaginous layer (Fig. 133). When seeds were imbibed at 35 °C for 24 h, there

Fig. 133. *Reboudia pinnata.* Mucilaginous seed adhering to soil surface by mucilaginous connections; ×46

was an acceleration in subsequent germination at the optimal temperature. After exposure to high temperatures of the two related Saharo-Arabian crucifers, the different germination rates may correlate with the possibility that *R. pinnata* originated in the summer rain area of the southern Sahara and Arabia region while *C. annua* originated in the northern Saharo-Arabian regions receiving winter rain (Gutterman 1990a).

The seeds of *Carrichtera annua* are mucilaginous and need light for germination. Seeds are dispersed by rain and adhere to the soil surface. If there is a long enough period of wetting, they may germinate in the same rain event in which they have been dispersed. High temperatures at the beginning of wetting impose thermo-inhibition. This inhibits germination in light and delays germination for a long time in darkness. At 15°C, *C. annua* seeds germinated within the first 24 h (to 72%) in light (CL) but only to 16.5% in the dark (CD). When imbibition at 15°C was followed by 24 h at 35°C (HL), germination of only 1% was observed in continuous light. After 30 days, germination reached 93% (CL), 90% (HL), 29% (CD), 0% (HD). There was no germination in the dark after exposure to high temperature, even after more than 50 days of imbibition. After 40 days of dry storage, rewetting brought only a few seeds to germination following another 15 days of imbibition. It seems that this mechanism may prevent germination of species from areas of winter rain after unexpected summer rain (Gutterman 1990a).

5.4.6 Species Habitat, Speed and Range of Temperatures for Germination

Germination of three geophyte species, *Tulipa systola, Bellevalia desertorum* and *B. eigii* (Liliaceae) inhabiting three different habitats near Sede Boker, was checked in different constant temperatures (Boeken and Gutterman 1990). *T. systola*, which inhabits the north-facing slopes, did not germinate at 25°C even after 56 days. This is not the case in *B. desertorum,* which inhabits south-facing slopes in this area and germinates well at 25°C. *B. eigii* inhabits the wadi between these slopes and has an intermediate response. Seeds of *B. desertorum* began to germinate first; seeds at 25°C began to germinate after 7 days of imbibition and reached ca. 60% germination after 50 days. Seeds of *B. eigii* started to germinate after ca. 20 days of imbibition at 25°C, but even after 56 days germination was still lower than 10%. However, at 20°C germination reached ca. 80% after 56 days. Seeds of *T. systola* at 20°C began to germinate only after 28 days' imbibition and, after 56 days, less than 50% had germinated. The more extreme the habitat of these plants, the earlier the onset of germination, the quicker the germination process and the higher the percentage of germination at higher temperatures.

A good correlation was found between temperatures suitable for germination of these species, the speed of germination and the drop in soil moisture after a rain, in each of the three micro-habitats. A difference of more than 10°C between the north- and south-facing habitats was found when measured at the time of germination (between 10−20 January 1989) in the natural

Temperatures 195

habitats. The largest difference was between the habitat of *T. systola*, on the north-facing slopes where the soil temperature measured less than 15 °C, the lowest temperature of the three habitats. At the other extreme, soil temperature in the habitat of *B. desertorum* on the south-facing slopes was over 25 °C at this time.

The quickest drop in soil moisture measured 12 days after a rain of 10.2 mm was found on the south-facing slopes. This was especially true of the top 1 cm of soil where seeds are situated when they germinate (Boeken and Gutterman 1990) (Fig. 134).

Correlation was also found in *Helianthemum vesicarium* and *H. ventosum*, between the range of temperatures, speed of germination and habitat direction and elevation (Sect. 5.2.2.2) (Gutterman and Agami 1987; Gutterman and Edine 1988).

5.4.7 Thermoperiodism Affecting Seed Germination

The germination of the seeds of many species is improved by a regime of varying diurnal temperatures compared with germination at constant temperatures (Harrington 1923). In the case of *Rumex obtusifolium* with seeds imbibed at 25 °C, a single exposure to extreme temperatures (44 °C for 1 – 3 h or to 4 °C for 32 h) increases germination from less than 10% to more than 70%. After a single exposure to a temperature of 32 °C for 3 h, only 10% germinated in comparison with 73% after exposure for 3 h to 44 °C. After a single exposure to 20 °C for 32 h, less than 10% germination was obtained in comparison with ca. 80% germination when exposed to 4 °C (Meneghini et al. 1968).

Seeds on the soil surface are exposed to the most extreme daily fluctuations in temperatures. The deeper a seed is buried in the soil, the less the difference between circadian fluctuations in temperatures. Fluctuations almost disappear at a depth of ca. 20 – 30 cm in the loess soils of the Negev highlands (Shachak 1975) (Sect. 4.1.3.1). It is possible that some species use this daily change in temperature to regulate germination from the optimum depth at the appropriate season of the year.

In *Emex spinosa* (Sect. 2.2.1) germination of non-leached aerial (54 – 60%) and subterranean (20 – 26%) propagules increased when imbibed at 15 or 20 °C for 18 h and then transferred for 8 h to 30 or 35 °C, compared with aerials (6 – 24%) and subterraneans (0 – 7%) in constant temperatures of 15 or 20 °C in the light and dark (Table 6). At constant temperatures, the optimal temperature for germination of *Emex spinosa* propagules is 20 – 25 °C (Evenari et al. 1977).

The highest germination (ca. 42%) of *Piptatherum miliaceum* (= *Oryzopsis miliacea*) intact (non acid-treated) seeds from populations of the northern Negev was observed in a combination of alternating temperatures (20/30 °C). In constant temperatures (20, 26 and 30 °C) as well as alternating 10/20 °C, germination did not exceed 15% after 14 days.

Three xerophytic species of the Chihuahuan desert of S.W. North America have been tested in a similar way. The highest germination was found in *Fou-*

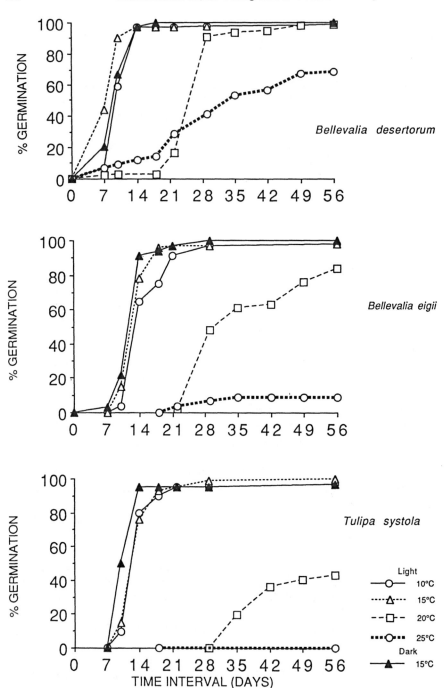

Fig. 134. *Bellevalia desertorum, B. eigii* and *Tulipa systola*: percentage of germination at 10, 15, 20 and 25 °C, in light or at 15 °C in dark, after 56 days of imbibition in water in petri dishes. (Boeken and Gutterman 1990)

Temperatures 197

quieria splendens under alternating temperatures of 20/40 °C, when the period of exposure at 40 °C was longer than 12 – 16 h. This species germinates in the Joshua Tree National Monument, California, after summer showers (Went 1948). The highest germination of *Agave lecheguilla* was found to occur after alternating temperatures of 20/35 °C. Under alternating 20/40 °C, there was a sharp decrease in germination even after 4 h in 40 °C (Freeman et al. 1977).

From 19 species of arid land grasses and shrubs collected mainly from populations in New Mexico, the highest degree of germination was obtained in 11 species in a diurnal range of temperatures. In another four, optimum germination occurred under either constant or a diurnal range of temperatures and, in five species, the highest percentage of germination was observed in constant temperatures (Tables 59, 60) (Sabo et al. 1979).

Seeds of *Carnegiea gigantea* (the Saguaro cactus of the Sonoran desert) will not germinate at constant temperatures of 15 °C and lower (0%), and very little above 35 °C (2%). The optimum constant temperature for germination is 25 °C (38%). Seeds will germinate in alternating temperatures of 35/25 °C (28%) and 30/20 °C (28%), but only in light. No germination occurred in the dark at all temperatures mentioned. Germination in alternating temperatures of 35/25 °C is of ecological importance as it enables this species to germinate in spite of the fact that daytime temperatures are higher than 35 °C (Alcorn and Kurtz 1959).

Of 12 species and populations of *Mimulus* of California, Texas, Arizona, Utah, Mexico, Argentina and Chile, seeds from Crystal Springs, California, at the lowest elevation (244 m), germinated in the widest range of alternating temperatures: 7/50 °C (5% germination), 21/50 °C (9%), 50/25 °C (12%). However, the highest percentage of germination occurred in 14/4 °C (100%); 17/9 °C (89%) and 9/21 °C (83%) also promoted good germination. The other

Table 59. The diurnal or constant temperatures recommended for optimum germination of nine species of grasses. (After Sabo et al. 1979)

	Species	Diurnal temperatures	Constant temperatures (°C)
A[a]	*Agropyron smithii*	18.5 °C (8 h) and 10 °C (16 h)	–
	Andropogon scoparius	27 °C (8 h) and 16.5 °C (16 h)	–
	Bouteloua gracilis	27 °C to 37 °C (8 h) and 31 °C (16 h)	–
	Distichlis stricta	40 °C (8 h) and 36 °C (16 h)	–
	Sporobolus contractus	15.5 °C to 19 °C (8 h) and 26.5 °C to 29.5 °C (16 h)	–
	Sporbolus cryptandrus	11 °C to 27 °C (8 h) and 35 °C to 37 °C (16 h)	–
B	*Bouteloua curtipendula*	12 °C (8 h) and 31 °C (16 h)	23
	Muhlenbergia wrightii	12 °C (8 h) to 34.5 °C (16 h)	12 to 34.5
C	*Hilaria Jamesii*	–	29

[a] A, Diurnal temperatures; B, constant and diurnal temperatures; C, constant temperatures.

Table 60. The diurnal or constant temperatures for optimum germination of ten species of shrubs. (After Sabo et al. 1979)

	Species	Circadian temperatures	Constant temperatures (%)
A[a]	*Cercocarpus montanus*	23 °C (8 h) and 11.5 to 15.5 °C (16 h)	–
	Chrysothamnus nauseosus ssp. *bigelovii*	20.5 °C (8 h) and 30 °C (16 h)	–
	Chrysothamnus nauseosus ssp. *consimilis*	13 °C (8 h) and 23 °C (16 h)	–
	Menodora scabra	24 °C (8 or 16 h) and 17 °C (16 or 8 h)	–
	Spheralcea incana	24 °C (8 h) and 17 °C to 20.5 °C (16 h)	–
B	*Artemisia frigida*	13 °C to 17 °C (8 h) and 23.5 °C (16 h)	17
	Atriplex confertifolia	16 °C (8 h) and 12 °C (16 h)	12
C	*Artemisia tridentata*	–	18.5
	Cowania stansburiana	–	25
	Fallugia paradoxa	–	22
	Sarcobatus vermiculatus	–	11

[a] A, Diurnal temperatures; B, constant and diurnal temperatures; C, constant temperatures.

species and populations from higher elevations did not germinate under such extreme temperatures (Vickery 1967).

5.5 Light and Germination

Light is one of the most important environmental signals in response to which the life cycles of plants and their seed germination are regulated. Depending on day length, many plants flower at the right time, sometimes in combination with temperatures. The same is true of seeds of various species which germinate at the right time of the year with a combination of optimum day length and temperatures. The light signal, including its wavelength intensity and day length, is also important for orientation of the seed with regard to the position of the micropile and depth of the seed in the soil. In *Pancratium maritimum* (Keren and Evenari 1974) and *Calligonum comosum* (Koller 1956) light penetrates to the embryo through the micropile when it is upwards if the seed is not too deep (Sect. 5.5.4.2). According to light, or lack of light, germination of seeds of various species is regulated. In others, germination may be inhibited. In many cases light acts in combination with temperature.

It is uncertain which are the signal or signals, light intensity, wavelengths or day length, which trigger a certain action. Many models have been tested to include as many results as possible, but none has given a complete explanation for all the known influences of light (Evenari 1965b; Bewley and Black

Light and Germination 199

1978, 1982; Mayer and Poljakoff-Mayber 1982; Fenner 1985). More recent models deal with the *dimeric mechanisms* of phytochrome action. They explain better (1) the very low fluence responses of light; (2) low fluence, as well as (3) the high irradiance responses to light (Van Der Woude 1989).

It is well known that the light sensitivity of seeds of many different species is affected, at least to a degree, by environmental factors and maternal influences when the seeds are still on the mother plant or in the fruit, even if the fruit has been disconnected from the mother plant. Light exerts many effects during the time between maturation and germination after seeds have imbided dew or very small amounts of rain for brief periods in combination with high or low temperatures. This can influence germination in a different way, as will be explained below.

5.5.1 Maturation in Natural Conditions Affecting Seed Light Sensitivity

In some plant species the amount of chlorophyll in the tissue surrounding the seed during maturation and dehydration of the seeds on the mother plant has an influence on the seed light sensitivity during germination. In species with a higher chlorophyll content, which acts as a far-red light filter during maturation and dehydration, there is also a higher percentage of light sensitive seeds (Cresswell and Grime 1981) (Sect. 2.4.8.2). There is a well-known canopy effect when green leaves act as a far-red light filter which affects the light sensitivity of seeds during the time of their maturation. This also prevents germination under the canopy in species with light-sensitive seeds (Black 1969; Vander Veen 1970; Fenner 1980a, b; Gorski 1975; King 1975; Cresswell and Grime 1981; Gutterman 1990a) (Sect. 2.4.8.2). One of the results of the mass germination of light-sensitive seeds after a fire is the absence of the green plant canopy which had earlier prevented germination of the seeds.

5.5.2 Maturation Under Artificial Light Affecting Seed Light Sensitivity

Environmental factors, such as light, which are involved during seed maturation, may also have an influence on seed light sensitivity. This is found in plants such as *Cucumis prophetarum* and *C. sativus*. During maturation, seeds from fruits stored under red or white light and imbibed in the dark germinated to a high level because of the high percentage of Pfr (the active form of phytochrome) in them. Germination was low in light, however, because of the low amount of photoreversible phytochrome in the seeds. The opposite is true after maturation of the seeds of this species, when fruit was stored in the dark or under far red light. Since the amount of photoreversible phytochrome is high, germination in light is high but low or very low when imbibition takes place in the dark, because the percentage of Pfr is very low (Gutterman and Porath 1975; Gutterman 1992d) (Sect. 2.4.8.1; Table 33; Figs. 46–51). Seed germination of *Arabidopsis thaliana* is affected by the R/FR ratio during maturation (McCullough and Shropshire 1970) (Sect. 2.4.8.1). Light treatment of *Portulaca oleracea* during seed maturation affects germination (Gutterman 1974) and also affects the germination of *Chenopodium album* (Karssen 1970) (Sect. 2.4.1).

5.5.3 Storage Conditions and Seed Light Sensitivity

Storage under conditions of high relative humidity and light increase the germination of Grand Rapid lettuce seeds in darkness from 21% at 15% relative humidity to 83% germination at 60–70% relative humidity (Koller 1972). The light requirement of the seed may be lost after imbibition under promotive light conditions and prevent germination by immediate dehydration of the seeds (Kincaid 1935). This has also been found in the case of *Piptatherum miliaceum* (= *Oryzopsis miliaceae*) (Poaceae) (Koller and Negbi 1959), a Mediterranean plant which likewise inhabits wadi beds in the deserts of Israel (Feinbrun-Dothan 1986). In other species, such as *Cucumis prophetarum, C. sativus* (Gutterman and Porath 1975) and *Gymnarrhena micrantha* (Koller and Roth 1964), the light effect disappears after dehydration and rehydration. Species such as *P. miliaceum* (Koller and Negbi 1959) and *Amaranthus retroflexus* (Amaranthaceae) (Kadman-Zahavi 1960), which also inhabit the Negev Desert highlands, retain their original sensitivity to light even after prolonged imbibition in darkness. Seed light sensitivity even increased in seeds of *Artemisia monosperma*, after 6–8 days of dark imbibition (Koller et al. 1964).

5.5.4 Types of Response of Seeds to Light

Different groups of responses to light have been found during the time of imbibition in the seeds of various desert species. There are seeds:

1. whose germination is accelerated by light:
   (a) in which red or white light accelerates but far red light inhibits germination, e.g. *Saguaro cactus* (*Carnegiea gigantea*) (Alcorn and Kurtz 1959) or *Portulaca oleracea* (Gutterman 1974);
   (b) in which the whole range of wavelengths of the visible spectrum increases germination in comparison with dark treatments, e.g. *Artemisia monosperma* (Koller et al. 1964; Koller 1969), *A. sieberi* (Evenari and Gutterman 1976).
2. whose germination is inhibited by light:
   (a) in all temperatures, e.g. *Pancratium maritimum* (Keren and Evenari 1974), *Bergeranthus scapiger* (Gutterman 1990a);
   (b) under low temperatures, continuous white light inhibits germination in comparison with short illuminations of white light, e.g. *Atriplex dimorphostegia* (Koller 1957), *Hyoscyamus desertorum* (Roth-Bejerano et al. 1971);

Light and Germination 201

(c) some species germinate in the dark although short periods of illumination do not interfere with the germination, because germination is only affected by darkness during the last stage of germination, e.g. *Pancratium maritimum* (Keren and Evenari 1974).
3. that are not light sensitive:
   (a) in the range of temperatures required for germination, e.g. *Blepharis* spp. (Table 45) (Sect. 3.3.3.4) (Gutterman 1972);
   (b) at low temperatures, but are so at high temperatures, e.g. *Lactuca serriola* (Small and Gutterman 1992b);
   (c) but become light sensitive after imbibition in supra-optimal temperatures, higher than the range of temperatures required for germination, e.g. *Carrichtera annua* (Gutterman 1990a).
4. whose germination is dependent on timing of dark and light periods:
   (a) timing of dark and light periods affects germination, e.g. *Plantago coronopus* (see below) (Koller and Negbi 1966);
   (b) photoperiodism affects seed germination, e.g. *Betula pubescens* (not a desert plant) (Black and Wareing 1955).

5.5.4.1 Seed Germination Accelerated by Light (Photoblastic Seeds)

Seeds that require light for germination only germinate when they are near or above the soil surface and imbibition continues for long enough to complete the process. Many seeds of desert plants that require light for germination are mucilaginous and adhere to the soil surface by wetting. After dispersal, the mucilaginous layer absorbs water and acts as a seed bed.

Desert plants whose seed germination is affected by light are as follows:

1. Seeds of *Portulaca oleracea* which show a typical red/far-red reaction. Red light for 30 min induces high germination (ca. 70%) and far-red light as much as in the dark (ca. 7% after 24 h of imbibition at 40 °C) (Gutterman 1974). This is similar to what has been found in Grand Rapids lettuce seeds at 26 °C (Evenari 1965b; Gutterman et al. 1972).
2. Seeds of desert plants that need light for germination, such as the mucilaginous achenes of *Artemisia monosperma*, in which all parts of the spectrum of the visible light accelerates germination in comparison with the dark (Sect. 5.3.2) (Koller et al. 1964; Koller 1969). Mucilaginous achenes of *A. sieberi* are dispersed in December/January. They adhere to the soil surface when the seed surface is wetted by dew or rain (Fig. 139). These seeds germinate after about 16 days of imbibition under light. As in *A. monosperma*, all parts of the spectrum of visible light accelerate germination in comparison with germination in dark (Sect. 5.3.2) (Evenari and Gutterman 1976).

5.5.4.2 Light-Inhibited Seed Germination (Negatively Photoblastic)

1. Germination regulated by lack of light is important to the survival of various desert plants, such as a few plants that inhabit sandy habitats and have been found to be negatively photoblastic. This means that light inhibits seed germination and the seeds require dark conditions and therefore need to be covered by sand in order to germinate. Examples are *Calligonum comosum* (see below) (Koller 1956), *Zygophyllum coccineum* (Zygophyllaceae) (Batanouny and Ziegler 1971), *Cakile maritima* (Brassicaceae) (Barbour 1970), *Aster tripolium* (Asteraceae), *Atriplex litoralis* (Chenopodiaceae) (Richter and Libbert 1967), *Spinifex hirsutus* (Harty and McDonald 1972), *Ipomea stolonifera* (Convolvulaceae) (A. Keren and M. Evenari, pers. comm.) and *Pancratium maritimum* (see below) (Keren and Evenari 1974).

In *P. maritimum* (Amaryllidaceae), the lower the light intensity after 14 days of imbibition, the higher is the percentage of germination. When seeds were exposed to light intensity of 1800 lx, germination was only 13%; in 52% light intensity 30% germinated; in 26% light intensity 54% germinated and, in complete darkness 96% germinated. It was found that seeds needed to be buried about 1.3 cm below the sand surface to reach the same degree of germination (94%) in light (8 – 10 lx) and in the dark. Germination of the control seeds in petri dishes in light or of seeds on the sand surface in light, reached only 14%. Seeds buried two-thirds in sand, but with the micropyle exposed to the light, germinated to 49%. If the micropyle was downwards, germination was 89%. In a range of temperatures from 15 to 30°C, interruption of 10 – 15 min of light every 48 h had no effect on germination, and the level of germination was the same as in continuous dark. This is because the inhibiting effect of light occurs only during the final stage of the germination process. No germination was observed between 35°C and 40°C (Keren and Evenari 1974).

Similar effects have been found in *Calligonum comosum* (Polygonaceae). This species is found in West Irano-Turanian and Saharo-Arabian regions and inhabits sandy areas in the Negev and Dead Sea valley. Germination reached 14.6% after day 7 of imbibition at 20°C in the light and 71.1% in the dark. At 26°C, germination reached 40% in light and 59.4% in the dark. At 30°C, germination reached only 0.4% in light and 12.0% in the dark. Light can only penetrate to the embryo when the dispersal unit is upright in the soil (36% germination). In any other position insufficient light penetrates to the embryo to inhibit germination and the percentage of germination is high (84%). This random mechanism spreads germination over time. More seeds germinate in following years when sand dunes move and the position of the seeds in the soil is changed and/or the seeds are covered with enough sand to prevent the light from reaching the embryos. At a certain depth, the rise in temperature is also limited. Light, high temperatures (30°C) and close contact with water are factors that inhibit germination of the seeds of this species (Koller 1956).

There are at least two ecologically important mechanisms in plants with negatively photoblastic seeds. First, seeds need to be buried in sand at a certain

Light and Germination 203

depth in order to germinate. This is important where the seeds cannot be dispersed by wind and depth allows them to germinate before the upper layer of sand dries out. The other advantage is that the seeds are randomly situated with their micropyles downwards, so that less light can reach the embryo. This means that, even under optimal conditions, only a portion of seeds will germinate. This is important for spreading risk. The downward orientation of the micropyle gives the root a better chance to penetrate into the sand before the upper layer dries.

Seeds of other species, e.g. *Artemisia monosperma*, which inhabit desert sandy areas, are light-sensitive. If they are too deep they will not germinate because of lack of light, and if they are too near the soil surface they will not germinate because the soil water content will become too low during the long period of imbibition that these seeds need for germination (10 days) (Sect. 4.1.3.2).

Seeds of species which need light for germination may be dispersed in deep cracks in the soil. The advantage of this is that they will not germinate while light is excluded. They cannot do so until they reach a position near the soil surface, either as a result of soil erosion or of the activity of animals in the area (Sects. 4.1.1, 4.1.2; Fig. 113a, b, c).

2. At temperatures as low as 15 or 20 °C the mucilaginous seeds of *Hyoscyamus desertorum* are almost completely inhibited by dark as well as by continuous irradiation by white light, but not by dark with short exposures to white light (10 min at 24-h intervals). From 25 °C to 35 °C prolonged white light and dark with short exposures to light no longer have an inhibiting effect. At these temperatures, germination in the dark is about 25%, whereas in continuous light and dark with short white light irradiation germination reaches 75% (Roth-Bejerano et al. 1971).

5.5.4.3 Seeds Not Light-Sensitive in the Range of Temperatures Required for Germination

1. Seed dispersal and germination of *Blepharis* spp. are regulated by the length of wetting and the amount of water. Different ecotypes and varieties that have been tested show that the range of temperatures that trigger germination may be from 8 to 40 °C, in light and dark, and germination occurs after only a few hours. Germination will be inhibited if there is an excess of water, because this prevents oxygen from penetrating the embryo through the heavy mucilaginous layer. Consequently, seeds can be dispersed by floods without suffering damage, and they germinate immediately after the excess water has disappeared (Gutterman et al. 1967; Witztum et al. 1969).

2. Seeds of *Lactuca serriola* are not sensitive to light at 15 °C but, from 20 to 30 °C, they are. No germination at all occurs in the dark. At 20 and 25 °C germination in light is almost 100%, while at 15 °C ca. 70% germinate in the dark and about 80% in the light (Small and Gutterman 1992b).

5.5.4.4 Timing of Rain, Dark and Light Affecting Germination

The seeds of some species germinate only in light, while others need dark. In the case of the mucilaginous seeds of *Plantago coronopus* (Plantaginaceae), the time at which rain begins (daytime or at night) has an influence on seed germination. The seeds of this species germinate under laboratory conditions to 60% if 8 h light follows 12 or even 16 h of darkness. When dark is followed by light, germination is low (Koller and Negbi 1966). If the same phenomenon occurs in natural habitats this means that rain at the beginning of the night may induce germination, but after a fall of rain during the day, germination may be poor.

5.5.4.5 Photoperiodism Affecting Seed Germination

A combination of the appropriate day length and range of temperatures may possibly ensure that seed germination occurs during the right season at times when adequate soil humidity is not the limiting factor. This is probably of special importance in deserts receiving rain during more than one season of the year. In areas which receive only winter rain, e.g. the Negev, such a combination of the appropriate day length, and suitable range of temperatures is very important for ensuring that seed germination will be prevented when it is too late in the season and there is no chance for the seedlings to develop into plants before the arrival of summer. This is in spite of the fact that sufficient water may be in the soil for germination to take place.

Until now, very few results illustrate the influence of day length on seed germination (Isikawa 1954, 1962; Black and Wareing 1955; Nagao et al. 1959; Koller 1972). Germination of the seeds of some desert species have been found to be affected by 'short days', e.g. *Atriplex dimorphostegia* (Koller 1972) and *Amaranthus retroflexus* (Kadman-Zahavi 1960), and of others by 'long days', e.g. *Artemisia monosperma*. This long day effect exists at the range of $10-25\,°C$ but disappears in temperatures of $30\,°C$ (Koller et al. 1964). Photoperiodism and temperature regulate the appearance of summer or winter populations of two genotypes of *Chenopodium album* in India (see Sect. 5.4.2) (Ramakrishnan and Kapoor 1974).

5.6 Annual Rhythm Regulating Seed Germination

An annual rhythm of germination has been found in the annual desert species *Mesembryanthemum nodiflorum*. Seeds will germinate during winter but not during summer, even when the mature capsules have been stored for several years under laboratory conditions. Groups of capsules of *M. nodiflorum*, collected after maturation in 1972 from an area between the Dead Sea and Jericho, were placed in germination conditions. During June and July 1976 almost no germination resulted. Germination experiments during which seeds were imbibed for 9 days at different temperatures were repeated at intervals during 1978. In April, germination reached 77–89%. In June and September it was very low (6–16%), but in December it was again high (66–81%) (Table 18A). Under temperatures alternating between 35 and 15°C, once in 24 h, it was observed that germination was 99% in April, 30% in June, 16% in September and 85% in December if the temperatures during the first 24-h period was 35°C. If the temperature during the first 24-h period was 15°C, the same trend was observed, but the percentage of germination was much lower (Table 18B) (Gutterman 1980/81a). During the summer of 1979 germination was again close to zero (results not shown) and in the winter of 1979/80 the percentages increased (Table 17).

Dry seeds in their capsules were taken to South Africa in 1987 where they were exposed to the hot summer for 5 months and were then brought back to Israel at the beginning of February 1988 (the middle of the winter in Israel). Germination tests were started 2 days after they had been brought back to Israel, and they germinated well, as is typical for the winter season. The months in which they were subjected to summer conditions in South Africa did not influence their germination behaviour (Gutterman 1990a).

5.7 Mass Germination and Seedling Emergence from Below the Soil Crust

Species such as *Mesembryanthemum nodiflorum* and *Aizoon hispanicum* produce very small seeds with a long period of afterripening. The seeds that are dispersed by rain (Gutterman 1982a, 1990b) into depressions are covered over the years and a hard soil crust develops above them. Mass germination of very large numbers of seeds in the same micro-habitat and at the same time has been observed after a heavy rain on many occasions. A very large number of seedlings rise together which enables them to break the soil crust and emerge from the surface of the soil. Emergence could not take place if only a single small seed germinated. Seeds of *Trigonella stellata, Plantago coronopus* and other atelechoric plants behave in a similar way after they have been covered. From the enormous numbers of seedlings that emerge, only a few dominant plants survive to produce seeds.

Once in several years, some of the annual species become dominant in their cover of very large areas. This phenomenon is very typical of the hot deserts of the world. It is the result of a combination of several environmental factors occurring during seed imbibition, which also correlate well with the 'readiness to germinate' of seeds in the seed bank of some species. This, in turn, could be the result of a combination of environmental and maternal influences on the seeds throughout their history; from mother plant development, seed maturation, seed dispersal, storage, seed rewetting and redrying, until the time of imbibition and germination (Gutterman 1983, 1986a). Mass germination occurs and, with suitable rain distribution, tremendous numbers of plants appear and cover large areas in the desert. The appearance of these species may be very rare during subsequent years (Gutterman 1969, 1983; Evenari and Gutterman 1976) (Sect. 6.5).

5.8 Conclusion

What is special about the survival mechanism of desert plants? In at least some of the species inhabiting more extreme deserts, there are combinations of survival mechanisms and strategies of seed dispersal and germination which ensure survival of the species. The production of large numbers of very small seeds, the history of each "seed" from, or even before, anthesis, storage conditions and the genetics of the species, affect the 'readiness for germination' at a particular rain event. Conditions during wetting and the duration of wetting by rain or flood, amounts and distribution of rain as well as microtopography, all affect germination. Inhibitors or salinity also affect germination and act as 'rain gauges'. There are also genetic influences which affect the time for germination (long- or short-term) and under what temperatures, for each species or ecotype. Temperature is the important regulator for the time of germination of winter and summer annuals in deserts with winter and summer rains. In these plants, the species germinating in summer and inhabiting areas with summer rain have different germination requirements than have related species which germinate in winter in areas receiving winter rain. Photoperiodism and thermoperiodism may also be important regulating mechanisms for seeds to germinate at the right time of the year and season.

There is relatively limited information on the different regulating and survival mechanisms of desert plants. More research is necessary for understanding better if and how seeds can predict the right time for germination.

A combination of several environmental factors which appear by chance and correlate with the "needs" of the seeds in the seed bank of some desert species, cause the mass germination which is so typical of the hot deserts of the world.

6 Germination, the Survival of Seedlings and Competition

6.1 Introduction

6.1.1 Selective Process of Seed Germination and Seedling Survival

Went (1953) has summarized his research carried out in the Mojave and Colorado deserts of North America (in collaboration with M. Juhren and E. Phillips) (Went 1948, 1949; Went and Westergaard 1949). He arrived at the conclusion that evolution plays only a minor role in the selection of desert annuals after they have germinated. 'The selective process must, therefore, operate during the germination stage' and the most important regulating factors are the amounts of rain and the temperature (Sects. 5.2.1, 5.4.1). This is because: (1) in most cases at least 50% of the seedlings that appear survive, flower and produce seeds, (2) in only relatively few cases do seedlings disappear after they have germinated and before flowering. This is usually due to consumption by animals. However, 'If these annuals germinate at the wrong temperature or after an insufficient rain, their seeds will not ripen and they will die out in that particular area' (Went 1953).

6.1.2 Rain Amount, Distribution and Survival

In general, the winter annuals of the Negev also have a high percentage of survival (43–67%), as an average of all the species that appear, survive and mature seeds, in a particular year on a northern facing hillslope in Avdat (Evenari and Gutterman 1976). The percentage of seedling survival is dependent upon the additional amounts and distribution of rain that appears during seedling development. In some species there is a causal relationship between the amount of rain and the number of seedlings that appear in a particular micro-habitat, as well as the number of seedlings of annual species that survive and mature seeds (Table 61).

Loria and Noy-Meir (1979/80) showed that on a loess desert plain at Sede Zin in the Negev (an area near Sede Boker with an average rainfall of ca. 100 mm (Sect. 1.1; Table 1)), the survival of 14 species of desert annuals increases very much in a good year, such as 1974, which had 155 mm precipitation. This is in comparison with a year with half the average annual precipitation (such as 1973 with 48 mm). The survival rate of *Filago desertorum* in-

Table 61. Amount of rainfall (mm), mean number of seedlings, number of surviving plants on which seeds matured, % of survivors ($\pm$ S.E.) of winter annuals excluding *Salsola inermis* and the total number of seedlings that appeared, under natural conditions over five seasons, from 1960 to 1965, on a north-facing hillslope near Avdat in the Negev desert. (After Evenari and Gutterman 1976)

Year	Rain (mm)	Mean total number of seedlings that appeared and survived								
		Winter annuals excluding *Salsola inermis*			*Salsola inermis*			Total number		
		Seedlings/ m^2	Surviving plants/m^2	Surviving plants (%)	Seedlings/ m^2	Surviving plants/m^2	Surviving plants (%)	Seedlings/ m^2	Surviving plants/m^2	Surviving plants (%)
1960/61	70.4	8.2$\pm$ 2.6	5.7$\pm$ 2.7	66.7$\pm$22.6	9.2$\pm$ 4.0	8.0$\pm$ 3.8	87.0$\pm$ 6.7	19.7$\pm$ 7.2	14.2$\pm$ 6.7	67.4$\pm$10.7
1961/62	64.7	18.2$\pm$ 4.4	7.5$\pm$ 1.7	43.3$\pm$ 8.0	450.0$\pm$184.0	52.7$\pm$49.4	12.2$\pm$ 7.0	476.0$\pm$190.0	62.7$\pm$51.0	16.6$\pm$ 8.4
1962/63	29.5	0	0	0	0	0	0	0	0	0
1963/64	165.0	129.0$\pm$25.0	67.1$\pm$11.1	60.0$\pm$ 8.6	44.8$\pm$ 21.6	12.7$\pm$ 3.5	57.4$\pm$10.2	364.0$\pm$ 46.0	256.0$\pm$58.0	65.5$\pm$ 7.3
1964/65	159.8	96.0$\pm$15.7	50.6$\pm$12.7	45.6$\pm$ 5.8	476.0$\pm$105.0	17.7$\pm$ 6.5	4.6$\pm$ 1.7	641.0$\pm$106.0	70.3$\pm$12.5	12.0$\pm$ 2.1

Introduction

creased from 24% in 1973 to 79% in 1974, of *Schismus arabicus* from 19% to 83%, of *Erodium oxyrhynchum* from 9% to 85% and of *Spergularia diandra* from 0 to 47%. The yield of seeds per m^2 also increased according to the amount of precipitation: in *F. desertorum* from 200 seeds/m^2 in 1973 to 12,000 in 1974; in *S. arabicus* from 128 to 1440, in *E. oxyrhynchum* from 0.4 to 196 and in *S. diandra* from 0 to 1140 seeds (Table 62).

Some species germinate after as little as 10 mm of rain. In many annual species, the more rain, the greater the number of seedlings that appear

Table 62. Number of seedlings, survival and seed production per m^2 of the main species average over all quadrats in the years 1973 with 48 mm of rain and 1974 with 155 mm of rain. Observed in the Negev desert on loess plain Sede Zin near Sede Boker. (After Loria and Noy-Meir 1979/80)

Plant species	No. of seedlings/m^2		Survival (%)		No. of seeds produced/m^2	
	1973	1974	1973	1974	1973	1974
Erodium bryoniifolium	0.9	15.3	9	85	0.4	196
Roemeria hybrida	0.04	2.5	0	65	0	984
Astagalus hispidulus	0.4	2.4	0	57	0	112
Spergularia diandra	5.4	9.7	0	47	0	1 140
Stipa capensis	5.9	2.1	1	81	0.2	40
Schismus arabicus	34.0	17.4	19	83	128.0	1 440
Filago desertorum	4.1	83.1	24	79	200.0	12 000
Gymnarrhena micrantha	28.8	12.0	11	48	4.8	720
Carrichtera annua	28.8	6.4	7	˙91	10.0	240
Trigonella stellata	8.2	13.0	6	55	10.0	280
Astragalus tribuloides	4.2	7.9	13	67	5.6	212
Plantago coronopus	8.4	4.7	22	81	40.0	136
Reboudia pinnata	2.6	0.7	8	82	2.0	112
Salsola inermis	27.4	4.9	2	23	2.0	164

Table 63. *Schismus arabicus* populations (per m^2) on a gradient of four dry habitats on the loess plain Sede Zin near Sede Boker in the Negev highlands, in 1973 and 1974. (After Loria and Noy-Meir 1979/80)

Year and rain	Habitat	I	II	III	IV
1973	Seedlings, 1st rain	334	–	7	–
48 mm	Seedlings, 2nd rain	121	115	2	–
	Mature plants	142	1	0.5	–
	Seedling survival, %	31.2	0.9	5.6	–
	Seeds/plant	13	10	24	–
	Seeds produced	1 800	10	12	–
1974	Seedlings	133	53	17	3.6
155 mm	Mature plants	112	50	13	3.4
	Seedling survival, %	84.2	94.3	76	94.4
	Seeds/plant	89	40	51	52.0
	Seeds produced	10 000	2000	658	178.0

(Tables 61–63) (Evenari and Gutterman 1976; Loria and Noy-Meir 1979/80). It is clear that the rain which causes seeds to germinate is not always sufficient for the establishment of the seedling and for completion of the life cycle. As already mentioned, in a winter with fewer rains the percentage survival of winter annuals is low, and in a good winter with quantities of rainfall above the annual average survival is much higher. This is provided that there is no inter- and intra-specific competition; as was found, for example, in *Salsola inermis* and *Diplotaxis harra* in 1964/65 (Table 64) (Sect. 6.2.1).

The greater the amount of rain that causes germination, the larger the number of seedlings that appear. When the additional amounts of rains are smaller, the amount of competition between plants is greater and, therefore, the percentage of seedling survival may be lower. Not only is the amount of rain important but also the amount of water that penetrates the soil in a particular micro-habitat (Table 63) (Loria and Noy-Meir 1979/80).

6.1.3 Micro-Habitat and Seedling Survival

On the loess plain Sede Zin near Sede Boker, it was observed in four dry micro-habitats that the more water that accumulated in a certain micro-habitat in one season, the more seedlings of *Schismus arabicus* appeared per m^2. In the year with 48 mm of rainfall (1973) the percentage of survival was low – 31% in habitat I, and very low in the two others, 5.6–0.9%. The following year (1974), when 155 mm of rain fell, the survival rate was 94–76%. During the later year the number of *S. arabicus* seedlings that appeared in these particular micro-habitats (I and II) was lower, but the number of seeds produced was much larger. In both years there was a large decrease in seed numbers produced per m^2 from habitats I to IV: from 1800 to 10 seeds were produced in 1973 after 48 mm of rain, and from 10,000 to 178 in 1974 after 155 mm (Table 63) (Loria and Noy-Meir 1979/80) (Sect. 3.1). Depressions and porcupine diggings were found to be unique micro-habitats in the Negev desert highlands (Sect. 6.3).

Some years there is more than one fall of rain per season which causes germination, up to a maximum of five such rain events. This occurred during the winters 1963/64 and 1979/80. The later the germination, the longer the day length in which plants develop with a maximum of 5 months between the first and last such rain (Fig. 135) (Chap. 1; Tables 3–5).

6.1.4 Day Length and Water Stress Affecting Life Span of Annuals

Deserts in which the beginning of the rainy season is very unpredictable, while the end is more predictable (Chap. 1), such as the Negev and other deserts nearby, are inhabited by at least three main types of annual species. They are grouped according to their responses to day length (Table 65) (Evenari and Gutterman 1966, 1985).

Table 64. Total and mean number of seedlings/m^2 in 16×1 m^2 squares (*a*) of *Artemisia sieberi* and *Diplotaxis harra* in the winter 1964/65 under natural conditions on a northern hill slope near Avdat. Number of surviving plants (*b*), number of plants which survived from preceeding season (*c*), number of plants of which survived until the end of the summer 1965 (*d*), number of plants which flowered at the end of the summer 1965 (*e*) ($\pm$S.E.). (After Evenari and Gutterman 1976)

	Artemisia sieberi				*Diplotaxis harra*				
	a	b	c	d	a	b	c	d	e
Total in 16 m^2	30	30	373	371	1020	0	709	656	346
Mean/m^2	1.9 ± 1.3	1.9 ± 1.3	23.3 ± 6.3	23.2 ± 6.3	63.7 ± 18.9	0	44.3 ± 6	41 ± 5.2	21.6 ± 5.6

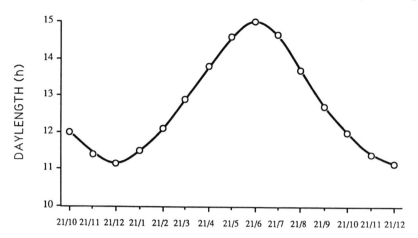
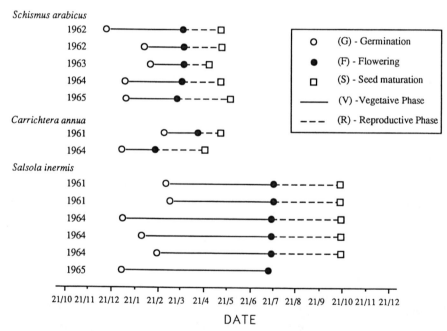

Fig. 135. Photoperiodic day length (from morning light intensity of 5 lx to the same intensity in the evening) and life cycle of three annuals: *Schismus arabicus*, *Carrichtera annua*, and *Salsola inermis*. As observed in Avdat over various years. G date of germination; F first appearance of flowers; S first seed maturation; V length of vegetative phase; R length of reproductive phase. (After Evenari and Gutterman 1966; Gutterman 1966, 1973)

Introduction 213

Table 65. The photoperiodic response of some desert plants as number of leaves at time of first flower bud appearance (±S.E.) and age (days) at time of first flower opening (±S.E.)

Group[a]	Plant species	Photoperiod[b]	No. of leaves at time of flower bud appearance	Age (days) at 1st flower opening
A	*Rumex cyprius*	S.g.	45.0±3.5	86.1±2.0
	(R. roseus)	L.g.	4.0±0	26.1±0.2
		S.o.	47.0±6.4	64.4±2.4
		L.o.	4.0±0	47.0±0
	Lappula spinocarpos	S.g.	27.4±2	64.6±2.3
	(Sclerocaryopsis	L.g.	6.0±0	23.0±0
	spinocarpos)	S.o.	20.5	62.3±1
		L.o.	7.3±0.8	45.3±1.1
B	*Carrichtera annua*	S.g.	17.4±1.39	65.7±3.4
		L.g.	5.3±0.31	31.0±2
		S.o.	6.0±0	45.0±0.53
		L.o.	6.0±0	43.0±0.49
	Reboudia pinnata	S.g.	18.0±0.76	75.0±4.3
		L.g.	10.5±0.81	48.3±3.6
		S.o.	7.6±0.66	57.9±0.74
		L.o.	7.6±0.87	57.7±1.58
C	*Salsola inermis*	S.g.	–	107.8±0.59
		L.g.	–	211.0±0
		S.o.	–	117.0±0
		L.o.	–	232.3±1.2
	Salsola volkensii	S.g.	–	103.7±0.18
		L.g.	–	158.0±0
		S.o.	–	129.0±0
		L.o.	–	231.7±15.0

[a] *A* Photoperiodic facultative LD plants; *B* day neutral; *C* photoperiodic facultative SD plants.
[b] *s.g.* Short day (8 h) in greenhouse; *L.g.* long day (20 h) in greenhouse; *S.o.* short day (8 h) outdoors; *L.o.* long day (20 h) outdoors.

The effect of the date of germination on the life span of annuals of the three types is summarized in Fig. 135, based on field observations carried out in Avdat in the Negev highlands 12 km from Sede Boker (Gutterman 1973; Evenari and Gutterman 1976).

6.1.4.1 Facultative Long-Day Plants

In facultative long-day plants, the later the germination in winter and the longer the day length, the shorter is the plant life span. There is also a reduction in the number of leaves at the appearance of the first flower bud and number of seeds as the plant is exposed to longer day length (Fig. 136). This day-length effect has been found in 12 Negev and annual species (of four families) out of 21 species tested under constant and different artificial day lengths: short days of 8 h or long days of 20 h (Table 65; Fig. 135) (Evenari and Gutterman

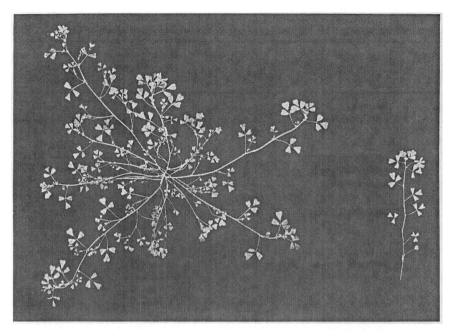

Fig. 136. Greenhouse-grown *Trigonella arabica* plants which flowered after 31 days under LD conditions (18 h, *right*) and after 94 days under SD conditions (9 h, *left*). (Gutterman 1989f)

1966, 1985; Gutterman 1982b). It has been observed in the Negev highlands that if germination occurs in natural habitats after a late rain in the season with longer days, the age of plants at flowering and seed maturation is much lower in winter annuals with a facultative long day response for flowering. Therefore, their chance of surviving and producing seeds before the dry and hot summer is higher (Fig. 135) (Evenari and Gutterman 1966; Gutterman 1973, 1989d, e).

6.1.4.2 Day-Neutral Plants

Day-neutral plants develop flowers shortly after germination in long or short days in outdoor conditions (Table 65). The longer the soil is wet, the higher the number of seeds produced and the longer the plant life span (Fig. 135) (Evenari and Gutterman 1966; Gutterman 1973, 1988b, 1989c, h).

6.1.4.3 Short-Day Facultative Plants

Short-day facultative plants, such as *Salsola inermis*, need a combination of short days and high temperatures for flowering (Table 65; Fig. 135) (Gutterman 1973, 1989b).

6.2 'Opportunistic' or 'Cautious' Strategy. Low or High Chance of Seedling Survival

According to the strategy of the regulation of seed dispersal (Sect. 3.6) and germination (Chaps. 2 and 5; Fig. 127a), some species can be divided into two extreme groups: (1) the plants with an 'opportunistic' strategy, in which the seedlings appear when there is high risk and therefore they have a low chance of survival if more rain does not follow that which engendered germination, and (2) the 'cautious' strategy in which seeds germinate when there is low risk and therefore a high chance of survival (see Chap. 5). In some species, both dispersal and germination mechanisms act together to regulate the number of seeds produced and time of their germination (Chaps. 2 and 5; Sect. 3.6) (Zohary 1937, 1962; Koller and Roth 1964; Gutterman et al. 1967; Evenari et al. 1977; Loria and Noy-Meir 1979/80; Ungar 1979; Venable and Lawler 1980; Venable 1985; Venable and Levin 1985).

6.2.1 Seedling Congestion and Survival

In species with the 'opportunistic' strategy (Fig. 122), very large numbers of seedlings may appear in the same micro-habitat and could be exposed to intraspecific competition. On many occasions it was found that the higher the number of seedlings of one species per unit area, the lower the number of seedlings that finish their life cycle successfully by producing seeds, or the fewer the seeds produced per plant.

Examples of this have been observed in the common bi-seasonal annual *Salsola inermis* of the Negev highlands at the Evenari Research Station, Avdat. These observations are summarized in Table 61 (Evenari and Gutterman 1976). In the winter of 1960/61, when the number of seedlings per m^2 was relatively low (about 20 per m^2), and in spite of the fact that there was only 70 mm of rainfall (near Avdat the annual average rainfall is 95 mm) (Evenari et al. 1982), 67% of the total seedlings survived and produced seeds. This included $9/m^2$ seedlings of *S. inermis* of which 87% survived. However, 1 year later, in the same plots, when 64.7 mm of rain fell and 476 seedlings per m^2 appeared, only 17% survived. This included $450/m^2$ seedlings of *S. inermis*, of which only 12% survived. In 1963/64 and 1964/65, when 160–165 mm of rain fell, almost the same phenomenon was repeated. In 1963/64, when 364 seedlings per m^2 appeared, 65.5% survived. This included 45 seedlings of *S. inermis*, of which 57% survived. In 1964/65, when 641 seedlings per m^2 appeared, only 12% survived. This included $476/m^2$ seedlings of *S. inermis* of which only 4.6% survived (Table 61). Thus, the more seeds of *S. inermis* that germinate beneath one dead mother plant, the fewer the number of seedlings that survive and the lower the yield of seeds from each plant (Evenari and Gutterman 1976).

In the same plots and during the years mentioned above, germination was noted of the perennial facultative plant *Diplotaxis harra*. During the rainy

Table 66. Coverage of *Artemisia sieberi* as a percentage of all the perennial shrubs, and the number of annual plant seedlings on the different slope faces. (After Gutterman and Herr 1981)

Hill slope orientation	*Artemisia* % coverage	Total number of seedlings
North	87	39
South	50	95
West	35	125
East	25	228

seasons of 1960/61 and 1961/62, when 70.4 and 64.7 mm of rain fell per season, respectively, an average of one to six seedlings per m^2 appeared and 50 to 38% survived. In 1962/63, a season with 29.5 mm of rain, no seedlings of *D. harra* appeared in the control areas. In 1963/64, a season with 165 mm of rain, $170/m^2$ seedlings appeared and 84% of them produced seeds. Of these, $44/m^2$ adult plants survived until 1964/65. In that season, 1020 seedlings appeared in an area of 16 m^2 ($64/m^2$) but none survived (Table 64).

Friedman and Orshan (1975) found that, in *Artemesia sieberi* (one of the dominant shrubs of the Irano-Turanian region of the Negev Desert highlands), there is mortality of seedlings beneath the adult plants, and even to a distance of 50 cm from their centres. The seedlings suffer heavy mortality during the summer. This is probably because of competition for water and an inhibitory effect. This inhibition is also true of germination and seedling establishment in other plant species. Gutterman and Herr (1981) found, on the hills near Sede Boker, that a lower percentage of *A. sieberi* plants in the vegetation resulted in higher germination and survival among seedlings (Table 66).

In some amphicarpic plants, the position of each of the two groups of seeds also correlated with a different strategy of dispersal (Chaps. 3, 5; Fig. 122), germination and seedling survival. The aerial achenes of *Gymnarrhena micrantha*, for instance, have an 'opportunistic' strategy. They germinate even when there is a low chance of survival. However, the subterranean achenes of the same plant have a 'cautious' strategy of germination. Survival of the seedlings, even in less favourable habitats, is always higher than that of the seedlings of the aerial achenes. Plants that originated from subterranean achenes not only produce more subterranean achenes but also many more aerial achenes than plants originating from aerial achenes. During the year with 48 mm of rain, only subterranean achenes were produced. This occurred in all habitats (Table 67) (Loria and Noy-Meir 1979/80). The 'cautious' strategy of the subterranean propagules not only gives a higher chance of survival by regulating the time of germination but, in addition, the seedlings are much more tolerant to water stress. They also germinate in situ where a mother plant has terminated its life cycle successfully in a more favourable micro-habitat (Koller and Roth 1964).

Blepharis spp., of the Negev, Sinai Peninsula and neighbouring desert areas, also have a 'cautious' strategy of seed dispersal. The number of seeds which

Table 67. *Gymnarrhena micrantha* populations (per m²) on a gradient composing four dry habitats in Sede Boker in 1973/1974. Seedlings germinating from aerial and subterranean achenes, and the production of aerial and subterranean achenes. (After Loria and Noy-Meir 1979/80)

Year and rain	Habitat	I		II		III		IV	
	Life cycle	Aerial	Sub.	Aerial	Sub.	Aerial	Sub.	Aerial	Sub.
1973 48 mm	Seedlings, 1st rain	2	—	—	—	3	0.5	—	—
	2nd rain	261	3	186	—	92	1	0.6	0.2
	Seedling survival, %	3.4	33.5	5.9	—	26.3	100	0	0
	Mature plants	9	1	11	—	25	1.5	—	—
	Seeds/plant: aerial	0	0	0	—	0	0	—	—
	Seeds/plant: subterr.	1	2	1	—	1	2	—	—
	Seeds produced: aerial	0	0	0	—	0	0	—	—
	Seeds produced: subterr.	9	2	11	—	25	3	—	—
1974 155 mm	Seedlings	86	26	15	19	18	8	1.8	10
	Mature plants	24	25	6	12	6.5	6	0.2	7
	Seedling survival, %	27.9	96.2	40	63.2	36.1	75	11.1	70
	Seeds/plant: aerial	2.1	117	0	96	1.9	259	0	210
	Seeds/plant: subterr.	1.5	2.4	1.3	2.2	1.5	3	2.0	2.7
	Seeds produced: aerial	50	2920	0	1150	12.5	1555	0	1472
	Seeds produced: subterr.	35	60	8	26	9.5	18	0.4	18.8

218 Germination, the Survival of Seedlings and Competition

are dispersed and the time of germination are regulated. These mechanisms ensure that only a portion of the seeds is released from the seed bank of the dead mother plant during any one rain event. The release of seeds is regulated by double safety mechanisms of two 'water clocks', and will occur only after enough rain has fallen to ensure the survival of the seedlings (Sect. 3.3.3.4) or after a flood in a wadi in which plants from previous years are situated (Gutterman et al. 1967).

The extremely hard seeds of some trees and shrubs that inhabit wadis in the Mojave and Colorado deserts of North America germinate after their seed coats have been washed and ground down by floods. Only thus is the embryo able to break through and germinate at the right time and in the right location (Went 1953).

The numerous annual species which have different mechanisms of seed dispersal and germination can be classified in three main groups: (A and B) those using one of the two strategies mentioned, (C) amphicarpic species in which the aerials have one strategy and the subterranians have another. There are also many species of annuals of the Negev Desert highlands which show, to a certain extent, a combination of these two extreme strategies (Fig. 122).

6.3 Depressions and Porcupine Diggings as Favourable Micro-Habitats in the Desert

When a rainfall of more than 3 mm/h occurs and falls for more than an hour, runoff water accumulates in depressions (Evenari et al. 1982; Danin 1983) and in porcupine diggings on all chalk and loess soils in the Negev Desert highlands.

Porcupines dig to obtain the subterranean organs of geophytes and hemicryptophytes (Gutterman 1982c, 1987, 1988a). The sizes of the holes depends on the depth of the organs consumed. They are about 30×10 cm in area and 5 to 25 cm deep. In some places the distribution of these holes may be as concentrated as one to even three per m^2. During summer these diggings form traps for organic matter, dispersal units, seed and soil particles. Runoff water, together with soil particles and seeds, accumulates during the rain events of the winter. Consequently, porcupine diggings provide uniquely favourable micro-habitats for desert annuals.

During their first year, less annuals germinate in the diggings than in the surrounding area. In later years, however, there is a succession of the seedlings of an increasing variety of species. This succession reaches its peak when about $50-60\%$ of the original digging has been covered. After this, there are still very good conditions for germination but less runoff water accumulates and competition results in the reduction of variety of species, biomass, surviving plants and seed yield (Fig. 137) (Gutterman 1989a; Gutterman et al. 1990). The succession of events in a porcupine digging is a typical system of disturbance in an ecological biome. Such a model was suggested by Tilman (1982) for plants and by Owen (1987) for animals. Porcupine diggings provide spaces

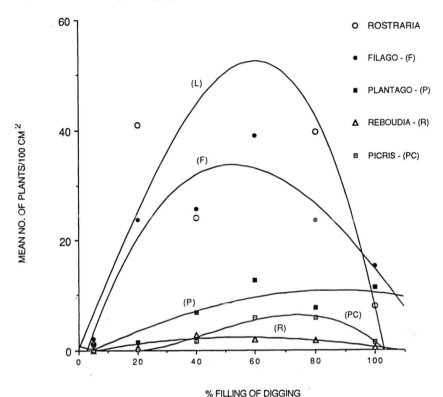

Fig. 137. Mean number of plants/100 cm^2 of *Rostraria smyrnaceae* (=*Lophochloa berythea*), *Filago desertorum*, *Picris cyanocarpa*, *Plantago coronopus* and *Reboudia pinnata* that appeared in 70 porcupine diggings in relation to the percentage of filling on a northern hill slope near Sede Boker. There was a significant increase in *R. smyrnaceae* between 40–60% filling and a significant decrease from this point to 100%. *R. symrnaceae* appeared in 42 of the 70 control areas with an average of 2.5±0.5 plants per control area. There was a significant increase of *P. cynocarpa* plants between 60–80% filling and a significant decrease from this point to 100%. *P. cyanocarpa* plants appeared in 14 of the 70 control areas with an average of 0.26±0.2 plants per control area. ○ *Rostraria* (*L*); ● Filago (*F*); ■ Plantago (*P*); △ Reboudia (*R*); ⊡ Picris (*PC*). (Gutterman et al. 1990)

for germination of annuals on the slopes, especially in areas where *Artemesia sieberi* is the dominant plant because of the allelopathic effect of adult plants of *A. sieberi* during germination and seedling development (Table 66) (Gutterman and Herr 1981). Shachak et al. (1991) found that different proportions of three species appeared in one particular year on a slope along a water-shed gradient. This composed 70% of the population of the annuals. These species, *Filago desertorum*, *Picris cyanocarpa* and *Bromus rubens*, appeared in 144 diggings on hill slopes. In 288 samples of porcupine diggings and matrix, 20,584 plants were identified: 18,542 of them in the diggings and 2,042 plants in the matrix. The three species showed different patterns of appearance along

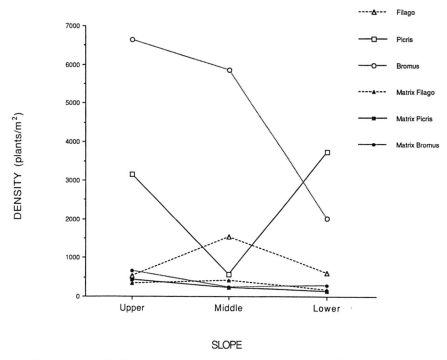

Fig. 138. Spatial distribution of *Filago desertorum*, *Picris cyanocarpa* and *Bromus rubens* in 144 porcupine diggings and 144 matrix samples along a slope. (Shachak et al. 1991)

the slope. *F. desertorum* was the only one whose abundance responded to the moisture gradient, with the peak in the middle of the slope. In contrast, the lowest numbers of *P. cyanocarpa* were observed in the middle of the slope. In *B. rubens*, the lower the slope the less the number of plants (Fig. 138) (Shachak et al. (1991)). According to Whittaker (1956, 1967), species abundance along a gradient forms a bell-shaped curve which is species-specific.

6.4 Drought Tolerance and the Survival of Seedlings

6.4.1 Seedling 'Point of No Return'

One very important survival mechanism in the desert is the drought tolerance of seedlings immediately after germination. *Salsola inermis* seedlings were found to be drought-tolerant until the radicle reached a certain size. Until this stage, even when they dried, they still survived and were able to renew themselves after rewetting. However, when they passed this 'point of no return', they did not survive after rewetting (Evenari et al. 1982). The same phenomenon was found by Friedman et al. (1981) in *Anastatica hierochuntica* seedlings but the 'point of no return' in this species is when the rootlets are 4–6 mm long. After dehydration for 1 week they renewed their growth within 8 h after wetting. In *A. hierochintica*, the 'point of no return' is reached when the roots are much longer than those of other desert plants observed.

6.4.2 Plant Drought Tolerance

In the amphicarpic annual *Gymnarrhena micrantha*, Koller and Roth (1964) found that there was a large difference between seedlings from aerial and subterranean achenes, in so far as the amount of water in the soil was concerned. The greater the water stress, the lower the survival of the seedlings from the aerial achenes; whereas subterraneans showed almost no mortality under these conditions. In containers in which irrigation was interruption every 7 days, all 24 seedlings from the aerial achenes died but only 9 of the subterranean achenes. Seedlings and adult plants from subterranean propagules of the amphicarpic *Emex spinosa* are more drought-resistant and flower earlier than their aerial counterparts (Evenari et al. 1977).

6.5 Mass Germination and Age Groups Replace Dead Plants

In the natural vegetation of the Negev highland the age groups of the *Zygophillum dumosum* shrubs were found by counting the numbers of annual rings in the stems (Evenari et al. 1982). From field observations and laboratory studies, it has been observed that perennial species such as *Artemesia sieberi* replace dead plants once in several years. In 1963/64, 165 mm of rain were received but during the period between 1 December to 2 January 120 mm accumulated in 10 days of well-distributed rain. Maturation of the seeds of *Artemesia sieberi* occurs during December and January, the middle of the rainy season. The seeds require light for germination. They adhere to the soil surface by the layer of mucilage which surrounds them when dispersed and wetted by rain or dew (Fig. 139). These seeds require about 16 days of imbibition before germination. This is why mass germination occurs only once in several years: such a long period of moist soil surface is very rare in the natural habitat (Tables 61, 64) (Evenari and Gutterman 1976).

Mass germination every 20–30 years is possibly the cause of the age groups in *Artemesia* populations. A few seedlings appear from time to time in other years, but do not survive for long (Evenari and Gutterman 1976). A similar phenomenon of age group distribution has been found among the common shrubs of the deserts of North America. The age groups of the Ocatillo (*Fouquieria splendens*), as well as of *Larrea tridentata*, are a result not only of lack of germination but also of low survival rates of the seedlings in most years (Freeman et al. 1977). (For mass germination of annual desert plants see Sect. 5.7.)

Fig. 139. Scanning electron microscope magnification of *Artemesia sieberi* achene adhering to the soil surface by its muculaginous seed coat. ×40

In the southern Namib extreme desert (the main Namib Sand Sea) the mean annual rainfall is less than 25 mm and unpredictable. After over 100 mm of rainfall in 1976 and again in 1978 the two endemic perennial species *Stipagrostis sabulicola* (Poaceae) and *Trianthema hereroensis* (Aizoaceae) germinated. After 11 years only 11% of *S. sabulicola* and none of the *T. hereroensis* plants had survived (Seely 1990).

6.6 Conclusion

There is insufficient information to answer questions such as: What are the influences of different amounts of rain on (1) the percentage of seeds from the seed bank that are 'ready for germination'; (2) the percentage of seedlings that appear in a particular micro-habitat from seeds that are 'ready for germination'; (3) seedling density and the reduction in survival through inter- and intra-specific competition; (4) seedling density, survival, and numbers of seeds matured; (5) seed dispersal and germination mechanisms affecting survival.

It seems that there must be at least some causal relationship between the amount of rain, the number of seedlings that appear, seedling density, the percentage of survival and seed yield. In many cases it has been observed that with more rain more seedlings appeared. More seedlings appeared after rain that had caused germination followed by less rain, while the higher seedling density,

Conclusion

the more competition resulted in less survival. Species which spread their germination in time overcome intra-specific, but not inter-specific, competition.

Survival, even after a very low rainfall, shows not only the ability of seeds to germinate, but also the ability of some desert species to complete their life cycle even when the percentage of survival is very low. In species whose seeds germinate in extreme desert, therefore, the development of seedlings must be very fast to enable them to finish their life cycle, even when no rain follows the precipitation that caused them to germinate. Day length and water stress seem to be important regulators of the flowering, seed maturation and life-cycle length which could increase the chance of producing seeds even after late germination. In seasons when the rainfall is above the annual average, many of the annuals produce large quantities of seeds which enlarge the seed bank for many years. Suitable conditions for the germination of seeds and the survival of seedlings to fill empty gaps in populations of perennial species are very rare, and occurs only once in several years.

In various species, different mechanisms achieve the same results in that only a very small percentage of the seed population germinates after one rain event. The others remain viable for the following rain, which spreads the risk. There are also different survival mechanisms that increase the chance of seedlings producing mature seeds.

7 Conclusion

The less it rains and the more unpredictable the amounts and dates of precipitation, the more extreme a desert. How can desert plants 'predict' when to germinate so that they may survive in such conditions? At what point does the plant switch from the seed, which is the phase of the life cycle most resistant to the environmental factors such as high temperatures, high salinity and dryness, to becoming a seedling, which is the most sensitive phase? It seems that different species develop various survival mechanisms and strategies which are followed through their life cycles. There are also environmental factors that have an influence on seed germinability: during seed maturation, dispersal, storage and the imbibition of water. These may affect the time and place of germination, as well as seedling survival, plant development, flowering, seed production and the termination of the life cycle at the right time.

As far as seed germinability is concerned, the fate of the next generation is in many species, at least to a certain degree, dependent on conditions of maturation while the seeds are still on the mother plant. Maternal position and environmental factors affect differences in germinability during development and seed maturation. In several species the last 5 – 15 days of maturation are critical. Seeds showing different germinability develop on the same mother plant, as do those on plants of the same species but growing in different environments.

One mother plant produces various seeds that differ in their germinability. They also differ, even on one branch and according to the position of the seeds in one fruit or dispersal unit. These phenomena act through different mechanisms in different plant species, but the result in many cases is that only a portion of the seed population will germinate after one rain event.

The biochemical events involved are still unknown. One can only speculate that during seed maturation, different factors affect the accumulation of different amounts of materials which are involved later in the germination process of the seeds. They may react through three main pathways: (1) they could lead to the development of seed coats with different degrees of impermeability to water, depending on day length and age effects; (2) these materials could result in an accumulation of germination inhibitors in the fruit, seeds and seed coats; (3) some of these materials may be enzymes and hormones which accumulate in the seed and/or embryo. It is probable that enzymes and other materials which accumulate in the seed, in different quantities depending on environmental conditions during maturation, affect seed germinability in different temperatures, thermoperiods or day length.

During the last stage of seed maturation day length affects the permeability to water of the seed coat in some species of the Fabaceae, and this effect is transferred from the leaves to the seeds in the covered fruit. It has been demonstrated in species of other families that day length effects are transferred from the leaves and affect the germinability of seeds from covered fruit. So far, neither the biochemistry of this process nor the material or materials which are transferred from the leaves to the seeds and thereby affect their degree of germinability have yet been identified.

In some species, the effect of day length on seed germination differs from its effect on flowering. It is, therefore, possible that, in these two mechanisms which appear in different stages of the life cycle, the regulation of seed germination and the regulation of flowering utilize two different biochemical pathways. Each of them regulates a particular stage at the appropriate time of the season and proper times to germinate and flower.

Seeds harvested from the same plant, at the same time and from the same environment, reached different levels of germination at various temperatures, according to their position and time of maturation. Seeds harvested from the same plants at different times and allowed to germinate at the same temperature, also reached different levels of germination. As genotypic inheritance increases the fitness of a species to its natural habitat so that its seeds germinate at the right time and in the right place, phenotypic influences result in an increase in diversity. This ensures the germination of only a portion of the seed population in one season. Other seeds remain in the seed bank, and germinate in the following season or seasons. In some of the plant species inhabiting the most extreme and unpredictable desert conditions, heteroblastic responses of seeds to maternal and environmental conditions contribute very strongly to survival. This prevents catastrophes that would otherwise be caused by mass germination after heavy rain, followed by a long dry period.

One of the most important questions about the pressure of seed predation and unpredictable environmental conditions is: What proportion of seeds produced by annuals remains viable in the following rainy season or seasons after the dry hot summer? This relates more to plants whose seeds are dispersed after maturation than to those whose seeds are not dispersed after maturation but are dispersed by wind or rains during the next season or seasons. In some species, seeds remain protected by the dead mother plant and germinate in situ or are dispersed later. Under extreme conditions mechanisms of seed dispersal by rain are more effective than immediate dispersal at the beginning of summer. Other synaptospermic mechanisms also protect seeds effectively until the following season with rain.

Nearly all the species whose seeds are dispersed during summer have large numbers of very small seeds. This spreads the risk of destruction. A few exceptions, however, produce larger and better-protected seeds.

In winter annual plants, whose seeds are dispersed by rain, maturation of the seeds takes place at the beginning of summer. The seeds are protected

Conclusion

throughout the summer by the dead parts of the mother plants. Some of them at least are dispersed during the rains of the next winter and more in the following winters. Seeds are dispersed mainly in December or January when most rain usually falls. The seeds of some species are dispersed during the following winter and adhere to the soil surface until they germinate, sometimes several years later. These seeds exchange one shelter, the capsule of the dead mother plant, for another – the soil crust. This occurs in species of *Mesembryanthemum, Aizoon,* etc. Seeds of plants such as *Blepharis* spp. are dispersed when rain falls, over a period of many years, a few seeds each winter or with each heavy rain. They are dispersed and germinate during the same rain event. Germination of such seeds, which are covered by a mucilaginous layer (myxospermy) and are dispersed by floods or runoff water, is inhibited until the excess water disappears.

Approximately 40 plant species inhabiting the deserts of Israel, whose seeds are dispersed by rain, are synoptospermic, at least during the first summer after seed maturation. The percentage of species of the arid Saharo-Arabian Hammada vegetation of the Negev with telechoric mechanisms is only 14.5%, while topochoric species comprise 85.5%. Some species of geophytes of the Negev highlands, whose seeds mature at the beginning of the summer, have been found to delay seed dispersal. Large numbers of seeds are still situated in the open capsules on the stem of the inflorescence at the beginning of the following winter. In this way, the seeds are protected from seed eaters during summer.

The ecological importance of desert annuals germinating season after season in the same depressions is obvious. Fitness increases as the gene pool decreases. Patchy distribution of local ecogenotypes therefore develops with time but at least three factors may counteract to lower it: (1) long distance pollination; (2) long distance seed dispersal; (3) the longer lived 'old' seeds of the seed bank germinate together with the 'new' seeds of the seed bank 'age structure'. All these may diminish the process of development of local ecogentotypes.

The subterranean 'seeds' of the amphicarpic annual plant species are not dispersed and germinate in situ where they mature, but the aerials are dispersed over long distances.

Seeds of the perennial *Artemisia sieberi,* one of the most common shrubs of the Negev, mature and the mucilaginous achenes are dispersed during December and January. Mass germination occurs only once in a number of years, which causes age groups to form. The same is also seen in another common Negev shrub, *Zygophyllum dumosum.*

When yearly fluctuations in dispersal and accumulation in the seed bank are summarized, they show that seeds of different species are dispersed during different seasons of the year. The most pronounced characteristic of the seeds of nearly all species of the Negev desert is their small size. Under pressure of seed consumption, an important means of ensuring survival is to produce many very small seeds. The few species with larger seeds, produce relatively small numbers, but these are well protected.

It is difficult to reach a conclusion as to whether there are strategies of seed dispersal which are typical of desert plants and at the same time do not exist in species that inhabit more humid areas. The mechanisms of dispersal and germination of seeds in species such as *Blepharis* spp. is one of the few examples of a sophisticated adaptation to completely unpredictable desert conditions. The subterranean 'seeds' of plants such as *Gymnarrhena micrantha*, which germinate in situ, have more 'cautious' strategies of germination and seedling survival than do aerial telechoric 'seeds'. These possess 'opportunistic' strategies which are also found in other species.

The time and storage conditions can be determined as the period and location in which the seeds are situated from the time of maturation until the time of germination. The success of the survival mechanisms of annual species in the extreme desert seems to be dependent upon the ability to ensure the existence of enough long-living seeds at the proper time and location to enable the continuation of the species, in particular locations.

Short-storage seeds, such as those of the East Sudanean epiphyte *Viscum cruciatum*, become attached to the beaks or legs of birds by the sticky fruit and are thus carried to the branches of *Acacia* trees. They germinate there after only a short time, as they do not require rain for germination. Seeds that become stuck to stones by the action of birds also germinate but the seedlings have no chance of survival.

Long-storage seeds of desert species mature in one season and may remain in storage throughout one or more unsuitable seasons before germination and seedling establishment takes place. Such seeds develop different germination regulating mechanisms. These include the need for periods of high temperature to overcome afterripening, or fluctuations of temperatures during summer to 'soften' the 'hard seeds'. A gradual increase in humidity enables free water to penetrate through the hilum valve of hard seeds. Scarification by floods of the hard coats of seeds of plants that grow in wadis, an annual cycle of dormancy and many other mechanisms ensure that at least some seeds will be 'ready for germination' at the right time and in the right place. Species that germinate in winter are inhibited by high temperatures at the time of wetting. This prevents germination after an unexpected summer rain. This mechanism may also prevent germination in summer of winter germinating plants in areas that receive winter and summer rains.

Some plants disperse their seeds at random over a number of years. Not more than 30% of *Blepharis* seeds, stored in the capsules on the dry inflorescence, are dispersed, even under optimal conditions, and the seeds are scattered at random. Seeds may remain viable on the dead mother plant for many years and some of them are dispersed from time to time. The longer the time of storage, the higher the percentage of capsules that are dispersed in one long rain event. The seeds of *Mesembryanthemum nodiflorum*, for example, remain attached to the soil surface after being dispersed by rain during the rainy season following their maturation. Each of the three groups of seeds, according to its position in the capsule, is released after a different amount of wetting, and is dispersed by drops of rain. The terminals are the first to be

Conclusion

dispersed and the basal seeds the last. The seeds of *Asteriscus pygmaeus* that are stored on the dead mother plant are also released by rain over many years, whorl after whorl, from the periphery to the centre.

Is the viability of seeds on the soil surface influenced by exposure to wetting by dew ca. 190 times per year, and, to a greater extent, by rainfall? Are such seeds affected by the high salinity of the desert soil surface? Is there a repair mechanism and do these factors prime the seed for germination? The answers may lead to a better understanding of one of the most fascinating events typical of deserts: which course of events affects the 'mass germination' and appearance of very large numbers of some species once over a period of many years.

In addition to soil moisture, cyclic changes in dormancy, etc., during imbibition there are at least two more environmental factors which could regulate seed germination at the right time to the right season, depending on the position of the seed in the soil: the thermoperiodic and photoperiodic regimes.

There appears to be at least some causal relationship between the amount of rain, the number of seedlings that appear, seedling density, the percentage of survival, and seed yield. In many cases it has been observed that after more rain more seedlings appeared. In some years more than one rain event triggers germination. When seedling density is higher, competition results in reduced survival. Species which spread their germination in time may, by so doing, overcome intra-specific competition, but not inter-specific competition.

Even after a very low rainfall, not only are seeds of some of the desert species able to germinate and finish their life cycle, even though the percentage of survival is very low. In species whose seeds germinate in extreme desert, the development of seedlings also needs to be very fast to enable them to finish their life cycle, even when no further rain follows the rain that triggered germination. Day length and water stress are important regulators of flowering, time of seed maturation and length of life cycle. They increase the chance of producing new seeds, even after late germination. In seasons when the rainfall is above the annual average, many annuals produce large quantities of seeds which enlarge the seed bank. This effect may last for many years. Suitable conditions for seed germination and survival of the seedlings of perennial species to fill empty gaps are very rare and occur once or twice over periods of many years. This explains of the age groups that are found in populations of some desert perennials.

One of the most important regulating mechanisms for survival in the unpredictable conditions of the extreme hot deserts is the germination of only a portion of long-living seeds at the right time and in the right location. This is controlled by genotypic inheritance as well as by phenotypic maternal and environmental factors during development, maturation, storage, and time required for imbibition. In addition, during the life cycles of annual plants, there are different survival mechanisms that increase the changes for seedlings to produce mature seeds even when they themselves germinate late in the season.

In addition to the all the above mentioned, one of the main questions to be answered is: what is the reason for the success of a plant such as *Schismus arabicus* in the unpredictable desert conditions? Is it a result of a combination of the "escape" strategy of seed dispersal and "opportunistic" strategy of germination? These plants disperse very small seeds in very large numbers which "escape" and avoid massive consumption. Even if the majority of the tremendous number of seeds are not successful in their "opportunistic" strategy of germination after about 10 mm of rain, still enough seedlings of the species survive to become a dominant annual species year after year in large areas and different habitats of the Negev Desert highlands. These surviving plants in their turn also disperse tremendous numbers of seeds. The very great number of seeds of *S. arabicus* and their many chances for only portions of seeds to germinate each season is possibly the successful combination of strategies for survival in this unpredictable extreme desert.

In contrast, in *Blepharis* spp., the result of the most sophisticated strategies of seed protection and "cautious" seed dispersal and germination is that relatively few plants are found in limited areas and habitats. The few, relatively large, seeds that are produced on this plant, and the relatively rare chances for dispersal and germination, is not as successful as the strategy of *S. arabicus* in these unpredictable desert conditions.

References

Abd el Rahman AA (1986) The deserts of the Arabian Peninsula. In: Evenari M, Noy-Meir I, Goodall DW (eds) Ecosystems of the world, 2B. Hot deserts and arid shrublands. B. Elsevier, Amsterdam, pp 29−54

Abulfatih HA (1983) Germination characteristics of nine species from Abha. J Arid Environ 6:247−251

Abramsky Z (1983) Experiments on seed predation by rodents and ants in the Israeli desert. Oecologia 57:328−332

Adams CA, Fjerstad MC, Rinne RW (1983) Characteristics of soybean seed maturation; necessity for slow dehydration. Crop Sci 23:265−267

Aitken Y (1939) The problem of hard seeds in subterranean clover. Proc R Soc Vict 51:187−210

Alcorn SM, Kurtz EB Jr (1959) Some factors affecting the germination of seed of the Saguaro cactus (*Carnegiea gigantea*). Am J Bot 46:526−529

Baitulin I, Rahimbaev I, Kamenetsky I (1986) Introduction and morphogenesis of wild *Allium* in Kazakhstan. Monograph, Nauka, Alma-Ata, Kazakhstan

Ballard LAT (1973) Physical barriers to germination. Seed Sci Tech 1:285−303

Ballard LAT (1976) Strophiolar water conduction in seeds of the Trifolieae induced by action on the testa at non-strophiolar sites. Aust J Plant Physiol 3:465−469

Bar Y, Abramsky Z, Gutterman Y (1984) Diet of gerbilline rodents in the Israeli desert. J Arid Environ 7:371−376

Barbour HG (1970) Germination and early growth of the strand plant *Cakile maritima*. Bull Torrey Bot Club 97:13−22

Barrett-Lennard RA, Gladstones JS (1964) Dormancy and hard-seededness in Western Australian serradella (*Ornithopus compressus* L.). Aust J Agric Res 15:895−904

Barton LV (1961) Seed preservation and longevity. Interscience, New York

Barton LV (1965 a) Seed dormancy: General survey of dormancy types in seeds, and dormancy imposed by external agents. In: Ruhland W (ed) Encyclopedia of plant physiology, XV. Differentiation and development, part 2. Springer, Berlin Göttingen Heidelberg, pp 699−720

Barton LV (1965 b) Dormancy in seeds imposed by the seed coat. In: Ruhland W (ed) Encyclopedia of Plant Physiology, XV. Differentiation and development, part 2. Springer, Berlin Göttingen Heidelberg, pp 727−745

Baskin JM, Baskin CC (1971 a) Germination of *Cyperus inflexus* Muhl. Bot Gaz 132:3−9

Baskin JM, Baskin CC (1971 b) The possible ecological significance of the light requirement for germination in *Cyperus inflexus*. Bull Torrey Bot Club 98:25−33

Baskin JM, Baskin CC (1976) Effects of photoperiod on germination of *Cyperus inflexus* seeds. Bot Gaz 137:269−273

Baskin JM, Baskin CC (1978) Seasonal changes in the germination response of *Cyperus inflexus* seeds to temperature and their ecological significance. Bot Gaz 139:231−235

Baskin JM, Baskin CC (1982) Effects of wetting and drying cycles on the germination of seeds of *Cyperus inflexus*. Ecology 63:248−252

Batanouny KH (1981) Ecology and flora of Qatar. University of Qatar. Alden, Oxford

Batanouny KH, Ziegler H (1971) Eco-physiological studies in desert plants. II. Germination of *Zygophyllum coccinum* seeds under different conditions. Oecologia 8:52−63

Beadle NCW (1952) Studies of halophytes. I. The germination of the seed and establishment of the seedlings of five species of *Atriplex* in Australia. Ecology 33:49−62

232 References

Beatley JC (1974) Phenological events and their environmental triggers in Mojave desert ecosystems. Ecology 55:856–863

Beattie AJ (1985) The evolutionary ecology of ant-plant mutualisms. Cambridge University Press, Cambridge

Beneke K (1991) Fruit polymorphism in ephemeral species of Namaqualand. M Sc Thesis, University of Pretoria

Beneke K, Van Rooyen MW, Theron GK, Van de Venter HA: Fruit polymorphism in ephemeral species of Namaqualand. III. Germination differences between the polymorphic diaspores. J Arid Environ (in press)

Berjak P, Farrant JM, Pammenter NW (1989) The basis of recalcitrant seed behaviour: cell biology of the homoiohydrous seed condition. In: Taylorson RB (ed) Recent advances in the development and germination of seeds. Plenum, New York, pp 89–108

Berkofsky L (1983) Rainfall patterns in the desert. World Meteorological Organization Technical Report WCP – 75

Bernstein RA (1974) Seasonal food abundance and foraging activity in some desert ants. Am Nat 108:490–498

Bewley JD (1980) Secondary dormancy (skotodormancy) in seeds of lettuce (*Lactuca sativa* cv. Grand Rapids) and its release by light, gibberellic acid and benzyladenine. Plant Physiol 49:277–280

Bewley JD, Black M (1978) Physiology and biochemistry of seeds in relation to germination, vol 1. Development, germination, and growth. Springer, Berlin Heidelberg New York

Bewley JD, Black M (1982) Physiology and biochemistry of seeds in relation to germination, vol 2. Viability dormancy and environmental control. Springer, Berlin Heidelberg New York

Bewley JD, Kermode AR, Misra S (1989) Desiccation and minimal drying treatments of seeds of Castor Bean and *Phaseolus vulgaris* which terminate development and promote germination cause changes in protein and messenger RNA synthesis. Ann Bot 63:3–17

Black M (1969) Light-controlled germination of seeds. Soc Exp Biol Symp 23:193

Black M, Naylor JM (1959) Prevention of onset of seed dormancy by gibberellic acid. Nature 184:468–469

Black M, Wareing PF (1955) Growth studies in woody species, VII. Photoperiodic control of germination in *Betula pubescens* Ehrh. Physiol Plant 8:300–316

Blair TA (1942) Climatology. Prentice-Hall, New York

Boeken B (1986) Utilization of reserves in some desert geophytes. Ph D Thesis, Ben-Gurion University of the Negev, Beer Sheva, Israel

Boeken B, Gutterman Y (1989a) *Bellevalia desertorum* and *B. eigii*. In: Halevy AH (ed) Handbook of flowering, vol VI. CRC, Boca Raton, pp 93–102

Boeken B, Gutterman Y (1989b) *Tulipa sistola*. In: Halevy AH (ed) Handbook of flowering, vol VI. CRC, Boca Raton, pp 648–653

Boeken B, Gutterman Y (1990) The effect of temperature on seed germination in three common bulbous plants of different habitats in the central Negev desert of Israel. J Arid Environ 18:175–184

Boeken B, Gutterman Y (1991) The effect of water on the phenology of the desert geophytes *Bellevalia desertorum* and *B. eigii*. Isr J Bot 40:17–31

Borthwick HA, Hendricks SB (1961) Effects of radiation on growth and development. In: Ruhland W (ed) Encyclopedia of Plant Physiology, XVI. External factors affecting growth and development. Springer, Berlin Göttingen Heidelberg, pp 299–330

Borthwick HA, Hendricks SB, Parker MW, Toole EH, Toole VK (1952) A reversible photoreaction controlling seed germination. Proc Natl Acad Sci 38:662–666

Borthwick HA, Hendricks SB, Toole EH, Toole VK (1954) Action of light on lettuce-seed germination. Bot Gaz 115:205–224

Brayton RD, Capon B (1980) Productivity, depletion, and natural selection of *Salvia columbariae* seeds. Aliso 9:581–587

Brown JH, Grover JJ, Davidson DW, Lieberman GA (1975) A preliminary study of seed predation in desert and montane habitats. Ecology 56:987–992

Brown JS, Venable DL (1986) Evolutionary ecology of seed-bank annuals in temporally varying environments. Am Nat 127:31–47

References

Burdett AN (1972) Antagonistic effects of high and low temperature pretreatments on the germination and pregermination ethylene synthesis of lettuce seeds. Plant Physiol 50:201–204

Burmil S (1972) Observations on the autoecology of *Carrichtera annua* L. Asch M Sc. Thesis Hebrew University of Jerusalem, Israel (in Hebrew)

Cantliffe DJ, Fiscer JM, Nell TA (1984) Mechanism of seed priming in circumventing thermodormancy in lettuce. Plant Physiol 75:290–294

Capon B, Brecht PE (1970) Variations in seed germination and morphology among populations of *Salvia columbariae* Benth. in southern California. Aliso 7:207–216

Capon B, Van Asdall W (1967) Heat pre-treatment as a means of increasing germination of desert annual seeds. Ecology 48:305–306

Capon B, Maxwell GL, Smith PH (1978) Germination responses to temperature pretreatment of seeds from ten populations of *Salvia columbariae* in the San Gabriel mountains and Mojave Desert, California. Aliso 9:365–373

Childs S, Goodall DW (1973) Seed reserves of desert soils. US/IBP Desert Biome Research Memo 73–75

Cloudsley-Thompson JL (1968) The Merkhiyat Jerbels: a desert community. In: Brown GW Jr (ed) Desert Biology, vol 1. Academic Press, New York, pp 1–20

Cloudsley-Thompson JL (1991) Ecophysiology of desert arthropods and reptiles. Springer, Berlin Heidelberg New York

Côme D, Corbineau F (1989) Some aspects of metabolic regulation of seed germination and dormancy. In: Taylorson RB (ed) Recent advances in the development and germination of seeds. Plenum, New York, pp 165–179

Comins NH, Hamilton WE, May RM (1980) Evolutionary stable dispersal strategies. J Theor Biol 82:205–230

Court D (1981) Succulent Flora of Southern Africa. A.A. Balkema, Cape Town

Courtney AD (1968) Seed dormancy and field emergence in *Polygonum aviculare*. J Appl Ecol 5:675–684

Cresswell EG, Grime JP (1981) Induction of a light requirement during seed development and its ecological consequences. Nature 291:583–585

Cumming BG (1963) The dependence of germination on photoperiod, light quality and temperature in *Chenopodium* spp. Can J Bot 41:1211

Dafni A, Shmida A, Avishai M (1981) Leafless autumn-flowering geophytes in the Mediterranean region – phytogeographical, ecological and evolutionary aspects. Plant Syst Evol 137:181–193

Danin A (1983) Desert vegetation of Israel and Sinai. Cana, Jerusalem

Danin A (1989) Nests of harvesting ants: a preferred primary habitat of wild *Beta Vulgaris* in Israel. In: Tan K (ed) The Davis & Hedge Festschrift. Edinburgh University Press, Edinburgh, pp 223–232

Danin A, Orshan G, Zohary M (1975) The vegetation of the northern Negev and the Judean desert of Israel. Isr J Bot 24:118–172

Danin A, Baker I, Baker HG (1978) Cytogeography and taxonomy of the *Portulaca oleracea* L. polyploid complex. Isr J Bot 27:177–211

Datta SC, Evenari M, Gutterman Y (1970) The heteroblasty of *Aegilops ovata* L. Isr J Bot 19:463–483

Datta SC, Evenari M, Gutterman Y (1972a) Photoperiodic and temperature responses of plants derived from the various heteroblastic caryopses of *Aegilops ovata* L. J Indian Bot Soc 50A:546–559

Datta SC, Gutterman Y, Evenari M (1972b) The influence of the origin of the mother plant on yield and germination of their caryopses in *Aegilops ovata* L. Planta 105:155–164

Davidson DW, Samson DA (1985) Granivory in the Chihuahuan Desert: interactions within and between trophic levels. Ecology 66:486–502

De Greef JA, Fredericq H, Rethy R, Dedonder A, De Petter E, Van Wiemeersch L (1989) Factors eliciting the germination of photoblastic Kalanchoe seeds. In: Taylorson RB (ed) Recent advances in the development and germination of seeds. Plenum, New York, pp 241–260

Dick-Peddie WA, Alberico MS (1977) Fire ecology study of the Chisos Mountains. Phase I. The Chihuahuan Desert Research Institute, Alpine, Texas

Do Cao T, Attims Y, Corbineau F, Côme D (1978) Germination des grains produits par les plantes de deux lignées d'*Oldenlandia corymbosa* L. (Rubiacees) cultivées dans des conditions contrôllées. Physiol Vég 16:521–531

Dornbos DL Jr, Mullen RE, Shibles RM (1989) Drought stress effects during seed fill on Soybean seed germination and vigor. Crop Sci 29:476–480

Dorne CJ (1981) Variation in seed germination inhibition of *Chenopodium bonus-henricus* in relation to altitude of plant growth. Can J Bot 59:1893

Duke SO, Egley GH, Reger BJ (1977) Model for variable light sensitivity in imbibed dark-dormant seeds. Plant Physiol 59:244–249

Egley GH (1989) Water-impermeable seed coverings as barriers to germination. In: Taylorson RB (ed) Recent Advances in the Development and Germination of Seeds. Plenum, New York, pp 207–224

Egley GH, Duke SO (1985) Physiology of weed seed dormancy and germination. In: Duke SO (ed) Weed physiology, vol I. Reproduction and ecophysiology. CRC, Boca Raton, Florida, pp 27–64

Egley GH, Paul RN Jr, Duke SO, Vaughn KC (1985) Peroxidase involvement in lignification in water-impermeable seed coats of weedy leguminous and malvaceous species. Plant Cell Environ 8:253

Ellner S, Shmida A (1981) Why are adaptations for long-range seed dispersal rare in desert plants? Oecologia 51:133–144

Ellner S, Shmida A (1990) An evolutionary game theory model for risk-taking. In: Ichishi T, Neyman A, Tauman LY (eds) Game theory and applications. Academic Press, London, pp 377–379

Engler A (1895) Über die Amphicarpie bei *Fleuryapodocarpa* Wedd. nebst einigen allgemeinen Bemerkungen über die Erscheinung der Amphicarpie und Geocarpie. Sitzungsber König Preuss Akad Wiss 5:1–10

Engler A (1964) Syllabus der Pflanzenfamilien, vol 2, 12th edn. Borntraeger, Berlin

Evenari M (1952) The germination of lettuce seeds. I. Light temperature and coumarin as germination factors. Palest J Bot 5:138–160

Evenari M (1961) Chemical influences of other plants (allelopathy). In: Ruhland W (ed) Encyclopedia of plant physiology, XVI. External factors affecting growth and development. Springer, Berlin Göttingen Heidelberg, pp 691–736

Evenari M (1963) Zur Keimungsökologie zweier Wüstenpflanzen. Mitt Florist-Soziol Arbeitsgem 10:70–81

Evenari M (1965a) Physiology of seed dormancy, after-ripening and germination. Proc Int Seed Test Assoc 30:49–71

Evenari M (1965b) Light and seed dormancy. In: Ruhland W (ed) Encyclopedia of plant physiology, XV. Differentiation and development, part 2. Springer, Berlin Heidelberg New York, pp 804–847

Evenari M (1981) Ecology of the Negev desert, a critical review of our knowledge. In: Shuval H (ed) Development in Arid Zone Ecology and Environmental Quality. Balaban ISS, Rehovot, pp 1–33

Evenari M, Gutterman Y (1966) The photoperiodic response of some desert plants. Z Pflanzenphysiol 54:7–27

Evenari M, Gutterman Y (1976) Observations on the secondary succession of three plant communities in the Negev desert, Israel. I *Artemisietum herbae albae*. In: Jacques R (ed) Hommage au Prof P Chouard. Etudes de Biologie Végétale. CNRS, Gif sur Yvette, Paris, pp 57–86

Evenari M, Gutterman Y (1985) Desert plants. In: Halevy AH (ed) Handbook on flowering. CRC, Boca Raton, pp 41–49

Evenari M, Koller D, Gutterman Y (1966) Effects of the environment of the mother plants on the germination by control of seed-coat permeability to water in *Ononis sicula* Guss. Aust J Biol Sci 19:1007–1016

Evenari M, Shanan L, Tadmor N (1971) The Negev, the challenge of a desert. Harvard University Press, Cambridge

Evenari M, Kadouri A, Gutterman Y (1977) Ecophysiological investigations on the amphicarpy of *Emex spinosa* (L.) CAMPD. Flora 166:223–238

References 235

Evenari M, Shanan L, Tadmor N (1982) The Negev. The challenge of a desert, 2nd edn. Harvard University Press, Cambridge, pp 437

Evenari M, Gutterman Y, Gavish E (1985) Botanical studies on coastal salinas and sabkhas of the Sinai. In: Friedman GM, Krumbein WE (eds) Hypersaline ecosystems. Ecological Studies 53. Springer, Berlin Heidelberg New York, pp 145–182

Fahn A (1947) Physico-anatomical investigations in the dispersal apparatus of some fruits. Palest J Bot 4:136–145

Fahn A (1967) Plant Anatomy. Pergamon, Oxford

Fahn A, Werker E (1972) Anatomical mechanisms of seed dispersal. In: Koslovsky TT (ed) Seed Biology vol I. Academic Press, New York, pp 151–221

Fahn A, Zohary M (1955) On the pericarpial structure of the legumen its evolution and relation to dehiscence. Phytomorphology 5:99–111

Feinbrun-Dothan N (1970) A key to the species of *Filago* L. sensu lato (Compositae) in Palestine. Isr J Bot 19:260–265

Feinbrun-Dothan N (1978) Flora Palestina, Part Three-Text. Israel Acad Sciences and Humanities, Jerusalem

Feinbrun-Dothan N (1986) Flora Palestina, Part Four-Text. Israel Acad Sciences and Humanities, Jerusalem

Feinbrun-Dothan N, Danin A (1991) Analytical flora of Eretz-Israel. Cana, Jerusalem

Felippe M, Dale JE (1968) Effects of CCC and gibberellic acid on the progeny of treated plants. Planta 80:344–348

Fenner M (1980a) The inhibition of germination of *Bidens pilosa* seeds by leaf canopy shade in some natural vegetation types. New Phytol 84:95–101

Fenner M (1980b) The induction of a light requirement in *Bidens pilosa* seeds by leaf canopy shade. New Phytol 84:103–106

Fenner M (1985) Seed ecology. Chapman and Hall, London

Fenner M (1991) The effects of the parent environment on seed germinability. Seed Sci Res 1:75–84

Freas KE, Kemp PR (1983) Some relationships between environmental reliability and seed dormancy in desert annual plants. J Ecol 71:211–217

Freeman CE, Tiffany RS, Reid WH (1977) Germination responses of *Agave lecheguilla, A. parryi*, and *Fouquieria splendens.* Southwest Nat 22:195–204

Friedman J, Orshan G (1975) The distribution emergence and survival of seedlings of *Artemisia herba alba* Asso in the Negev desert of Israel in relation to distance from the adult plants. J Ecol 63:627–632

Friedman J, Stein Z (1980) The influence of seed dispersal mechanisms on the dispersion of *Anastatica hierochuntica* (Cruciferae) in the Negev desert Israel. J Ecol 68:43–50

Friedman J, Gunderman N, Ellis M (1978) Water response of the hygrochastic skeletons of the true rose of Jericho (*Anastatica hierochuntica* L.). Oecologia 32:289–302

Friedman J, Stein Z, Rushkin E (1981) Drought tolerance of germinating seeds and young seedlings of *Anastatica hieronchuntica* L. Oecologia 51:400–403

Galil J (1938) The germination of *Viscum crucatum* Sieb. Palest J Bot 1:103–105

Gill AM, Cheney NP, Walker J, Tunstall BR (1986) Bark losses form two eucalypt species following fires of different intensities. Aust For Res 16:1–7

Gorski T (1975) Germination of seeds in the shadow of plants. Physiol Plant 34:342

Green E (1991) Transplant experiments with several Mediterranean plants which also occur in the Negev highlands. M Sc Thesis, Hebrew University of Jerusalem, Israel (in Hebrew)

Grey D, Thomas TH (1982) Seed germination and seedling emergence as influenced by the position of development of the seed on, and chemical applications to, the parent plant. In: Khan AA (ed) The physiology and biochemistry of seed development, dormancy and germination. Elsevier, New York, pp 81–110

Griffiths JF (ed) (1972) Climates of Africa, world survey of climatology, vol 10. Elsevier, Amsterdam, pp 1–604

Griffiths JM, Driscoll DM (1982) Survey of climatology. Chas E Merrill, Columbus, Ohio

Gupta RK (1986) The Thar desert. In: Evenari M, Noy-Meir I, Goodall DW (eds) Ecosystems of the world, 12B. Hot deserts and arid shrublands B. Elsevier, Amsterdam, pp 55–100

236 References

Gutterman Y (1966) The photoperiodic response of some desert plants and the effect of the environment of the mother plants on the germination by control of seed coat permeability to water in *Ononsis sicula* Guss. M Sc Thesis, The Hebrew University of Jerusalem (in Hebrew)

Gutterman Y (1969) The photoperiodic response of some plants and the effect of the environment of the mother plants on the germination of their seeds. Ph D Thesis, The Hebrew University of Jerusalem (in Hebrew with English summary)

Gutterman Y (1972) Delayed seed dispersal and rapid germination as survival mechanisms of the desert plant *Blepharis persica* (Burm.) Kuntze. Oecologia 10:145 – 149

Gutterman Y (1973) Differences in the progeny due to daylength and hormonal treatment of the mother plant. In: Heydecker W (ed) Seed Ecology. Butterworths, London, pp 59 – 80

Gutterman Y (1974) The influence of the photoperiodic regime and red-far red light treatments of *Portulaca oleracea* L. plants on the germinability of their seeds. Oecologia 17:27 – 38

Gutterman Y (1978a) Seed coat permeability as a function of photoperiodical treatments of the mother plant during seed maturation in the desert annual plant: *Trigonella arabica* del. J Arid Environ 1:141 – 144

Gutterman Y (1978b) Germinability of seeds as a function of the maternal environments. Acta Hortic 83:49 – 55

Gutterman Y (1978c) Influence of environmental conditions and hormonal treatment of the mother plants during seed maturation on the germination of their seeds. In: Malik CP (ed) Advances in plant reproductive physiology. Kalyani, New Delhi, pp 288 – 294

Gutterman Y (1980/81a) Annual rhythm and position effect in the germinability of *Mesembryanthemum nodiflorum*. Isr J Bot 29:93 – 97

Gutterman Y (1980/81b) Review: Influences on seed germinability: phenotypic maternal effects during seed maturation. In: Mayer AM (ed) Control mechanisms in seed germination. Isr J Bot 29:105 – 117

Gutterman Y (1981) Influence of quantity and date of rain on the dispersal and germination mechanisms, phenology, development and seed germinability in desert annual plants, and on the life cycle of geophytes and hemicryptophytes in the Negev desert. In: Shuval H (ed) Developments in arid zone ecology and environmental quality. Balaban ISS, Philadelphia, pp 35 – 42

Gutterman Y (1982a) Survival mechanisms of desert winter annual plants in the Negev highlands of Israel. In: Mann HS (ed) Scientific reviews on arid zone research I. Scientific Publishers, Jodhpur, India, pp 249 – 283

Gutterman Y (1982b) Phenotypic maternal effect of photoperiod on seed germination. In: Khan AA (ed) The physiology and biochemistry of seed development, dormancy and germination. Elsevier, Amsterdam, pp 67 – 79

Gutterman Y (1982c) Observations on the feeding habits of the Indian crested porcupine (*Hystrix indica*) and the distribution of some hemicryptophytes and geophytes in the Negev desert highlands. J Arid Environ 5:261 – 268

Gutterman Y (1983) Mass germination of plants under desert conditions. Effects of environmental factors during seed maturation, dispersal, germination and establishment of desert annual and perennial plants in the Negev Highlands, Israel. In: Shuval HI (ed) Developments in ecology and environmental quality. Balaban ISS, Rehovot, pp 1 – 10

Gutterman Y (1985) Flowering, seed development and the influences during seed maturation on seed germination of annual weeds. In: Duke SD (ed) Weed physiology I. CRC, Boca Raton, pp 1 – 25

Gutterman Y (1986a) Influences of environmental factors on germination and plant establishment in the Negev Desert Highlands of Israel. In: Jass PJ, Lynch PW, Williams OB (eds) Rangelands: a resource under siege. Aust Acad Sci, Canberra, Australia, pp 441 – 443

Gutterman Y (1986b) Are plants which germinate and develop during winter in the Negev Desert highlands of Israel, winter annuals? In: Dubinsky Z, Steinberger Y (eds) Environmental quality and ecosystem stability, vol III/A. Bar-Ilan University Press, Ramat-Gan Israel, pp 135 – 144

Gutterman Y (1987) Dynamics of porcupine (*Hystrix indica* Kerr) diggings; their role in the survival and renewal of geophytes and hemicryptophytes in the Negev desert highlands. Isr J Bot 36:133 – 143

References 237

Gutterman Y (1988a) An ecological assessment of porcupine activity in a desert biome. In: Ghosh PK, Prakash I (eds) Ecophysiology of desert vertebrates. Scientific Publishers, Jodhpur, India, pp 289–372

Gutterman Y (1988b) Day neutral flowering in some desert *Blepharis* species. J Arid Environ 12:215–221

Gutterman Y (1989a) The importance of the porcupine (*Hystrix indica* Kerr) diggings as microhabitats for the renewal and survival of annual desert plants in some of the sites of the Negev desert highlands, Israel. In: Steinberger Y, Loria M (eds) Environmental quality and ecosystem stability, IV-B. ISEEQS Publ, Jerusalem, pp 413–421

Gutterman Y (1989b) *Salsola inermis* and *S. volkensii*. In: Halevy AH (ed) Handbook of flowering, vol VI. CRC, Boca Raton, pp 553–557

Gutterman Y (1989c) *Carrichtera annua*. In: Halevy AH (ed) Handbook of flowering, vol VI. CRC, Boca Raton, pp 157–161

Gutterman Y (1989d) *Gymnarrhena micrantha*. In: Halevy AH (ed) Handbook of flowering, vol VI. CRC, Boca Raton, pp 356–359

Gutterman Y (1989e) *Schismus arabicus*. In: Halevy AH (ed) Handbook of flowering, vol VI. CRC, Boca Raton, pp 564–568

Gutterman Y (1989f) *Trigonella arabica* and *T. stellata*. In: Halevy AH (ed) Handbook of flowering, vol VI. CRC, Boca Raton, pp 641–647

Gutterman Y (1989g) *Colchicum tunicatum*. In: Halevy AH (ed) Handbook of flowering, vol VI. CRC, Boca Raton, pp 234–242

Gutterman Y (1989h) *Blepharis* sp. In: Halevy AH (ed) Handbook of flowering, vol VI. CRC, Boca Raton, pp 108–116

Gutterman Y (1990a) Do the germination mechanisms differ in plants originating in deserts receiving winter or summer rain? In: Mayer AM (ed) Special edition in memory of the late Prof. M. Evenari. Isr J Bot 39:355–372

Gutterman Y (1990b) Seed dispersal by rain (ombrohydrochory) in some of the flowering desert plants in the deserts of Israel and the Sinai Peninsula. Mitt Inst Allg Bot Hamb 23b:841–852

Gutterman Y (1991) Comparative germination study on seeds matured during winter or summer of some bi-seasonal flowering perennial desert plants from the Aiozaceae. J Arid Environ 21:283–291

Gutterman Y (1992a) Ecophysiology of Negev Upland annual grasses. In: Chapman GP (ed) Desertified grasslands: their biology and management. Linn Soc Symp Ser, Academic Press, London, pp 145–162

Gutterman Y (1992b) Maternal effects on seeds during development. In: Fenner M (ed) Seeds: the ecology of regeneration in plant communities. CAB International UK, pp 145–162

Gutterman Y (1992c) Maturation dates affecting the germinability of *Lactuca serriola* L. achenes collected from a natural population in the Negev desert highlands. Germination under constant temperatures. J Arid Environ 22:353–362

Gutterman Y (1992d) Influences of daylength and red or far red light during the storage of post harvested ripe *Cucumis prophetarum* fruit, on the light germination of the seeds. J Arid Environ 23:443–449

Gutterman Y: Germinability under natural temperatures of *Lactuca serriola* L. achenes matured and collected on different dates from a natural population in the Negev desert highlands. J Arid Environ (in press)

Gutterman Y, Agami M (1987) A comparative study of seeds of *Helianthemum vesicarum* Boiss. and *H. ventosum* Boiss. perennial desert shrub species inhabiting two different neighbouring habitats in the Negev desert highlands, Israel. J Arid Environ 12:215–221

Gutterman Y, Boeken B (1988) Flowering affected by daylength and temperature in the leafless-flowering desert geophyte *Colchicum tunicatum* its life cycle and vegetative propagation. Bot Gaz 149:382–390

Gutterman Y, Edine L (1988) Variations in seed germination of *Helianthemum vesicarium* and *H. ventosum* from populations of two different altitudes in the Negev highlands, Israel. J Arid Environ 15:261–262

Gutterman Y, Evenari M (1972) The influence of day length on seed coat colour an index of water permeability of the desert annual *Ononis sicula* Guss. J Ecol 60:713–719

Gutterman Y, Ginott S: The long-term protected "seed bank" in the dry inflorescents the mechanism of achenes (seeds) dispersal by rain (ombrohydrochory) and the germination of the annual desert plant *Asteriscus pygmaeus*. J Arid Environ (in press)

Gutterman Y, Herr N (1981) Influences of porcupine (*Hystrix indica*) activity on the slopes of the northern Negev mountain – germination and vegetation renewal in different geomorphological types and slope directions. Oecologia 51:332–334

Gutterman Y, Heydecker W (1973) Studies of the surfaces of desert plant seeds, I Effect of day length upon maturation on the seed coat of *Ononis sicula* Guss. Ann Bot 37:1049–1050

Gutterman Y, Porath D (1975) Influences of photoperiodism and light treatments during fruits storage on the phytochrome and on the germination of *Cucumis prophetarum* L. and *Cucumis sativus* L. seeds. Oecologia 18:37–43

Gutterman Y, Witztum A (1977) The movement of integumentary hairs in *Blepharis ciliaris* L. Burtt. Bot Gaz 138:29–34

Gutterman Y, Witztum A, Evenari M (1967) Seed dispersal and germination in *Blepharis persica* (Burm) Kuntze. Isr J Bot 16:213–234

Gutterman Y, Witztum A, Evenari M (1969a) Physiological and morphological differences between populations of *Blepharis persica* (Burm) Kuntze. Isr J Bot 18:89–95

Gutterman Y, Witztum A, Evenari M (1969b) Rain clocks in *Blepharis* plants. Mada 13:276–281

Gutterman Y, Evenari M, Heydecker W (1972) Phytochrome and temperature relations in *Lactuca sativa* L. Grand rapids seed germination after thermo-dormancy. Nat New Biol 235:144–145

Gutterman Y, Thomas TH, Heydecker W (1975) Effect on the progeny of applying different day length and hormone treatments to parent plants of *Lactuca scariola*. Physiol Plant 34:30–38

Gutterman Y, Evenari M, Cooper R, Levy EC, Lavie D (1980) Germination inhibition activity of a naturally occurring lignan from *Aegilops ovata* L. in green and infrared light. Experientia 26:662–663

Gutterman Y, Evenari M, Went FW (1982) The dynamics of seed populations in the soil in the desert area of Israel. Final Report (3rd year) to the US-Israel Binational Science Foundation

Gutterman Y, Golan T, Garsani M (1990) Porcupine diggings as a unique ecological system in a desert environment. Oecologia 85:122–127

Hagon MW, Ballard LAT (1970) Reversibility of strophiolar permeability to water in seeds of subterranean clover (*Trigonium subterraneum* L.). Aust J Biol Sci 23:519–528

Halevy G (1974) Effects of gazelles and seed beetles (Bruchidae) on germination and establishment of *Acadia* species. Isr J Bot 23:126–192

Hamilton WD, May RM (1977) Dispersal in stable habitats. Nature 269:578–581

Hamly DH (1932) Softening of the seeds of *Melilotus alba*. Bot Gaz 93:345–375

Harper JL (1977) Population biology of plants. Academic Press, London

Harrington GT (1923) Use of alternating temperatures in the germination of seeds. J Agric Res 23:295

Harrington JF (1963) Practical instructions and advice on seed storage. Proc Int Seed Test Assoc 28:989–994

Harrington JF (1972) Seed storage and longevity. Seed Biol 3:145–245

Harty RL, McDonald TJ (1972) Germination behaviour in beach spinifex (*Spinifex hirsutus* Labill.). Aust J Bot 20:241–253

Hegarty TW (1973) Temperature relations of germination in the field. In: Heydecker W (ed) Seed ecology. Butterworth, London, pp 411–431

Hegarty TW (1978) The physiology of seed hydration and dehydration, and the relation between water stress and the control of germination: a review. Plant Cell Environ 1:101–119

Heide OM, Junttila O, Samuelsen RT (1976) Seed germination and bolting in red beet as affected by parent plant environment. Physiol Plant 36:343–349

Heithaus ER (1981) Seed predation by rodents on three ant-dispersed plants. Ecology 62:136–145

Herre H (1971) The genera of Mesembryanthemaceae. Tafelberg Uitgewers Beperk, Cape Town

Heydecker W (1977) Stress and seed germination: an agronomic view. In: Khan AA (ed) The physiology and biochemistry of seed dormancy and germination. North Holland, Amsterdam, pp 237–282

Heydecker W, Higgins J, Gulliver RL (1973) Accelerated germination by osmotic seed treatment. Nature 246:42

References

Heyn CC, Dagan O, Nachman B (1974) The annual *Calendula* species: taxonomy and relationships. Isr J Bot 23:169–201

Hilhorst HWM, Karssen CM (1989) The role of light and nitrate in seed germination. In: Taylorson RB (ed) Recent advances in the development and germination of seeds. Plenum, New York, pp 191–205

Holm LG, Plucknett DL, Pancho JV, Herberger JP (1977) The world's worst weeds: distribution and biology. University Press of Hawaii, Honolulu

Hord O (1986) Foraging behaviour of three species of harvesting ants in Avdat. M Sc Thesis, Hebrew University of Jerusalem (in Hebrew with English Abstr)

Hyde EOC (1954) The function of the hilum in some Papilionaceae in relation to the ripening of the seed and the permeability of the testa. Ann Bot 18:241–257

Ihlenfeldt H-D (1983) Dispersal of Mesembryanthemaceae in arid habitats. Sonderb Naturwiss Ver Hamb 7:381–390

Ingram TJ, Browning G (1979) Influence of photoperiod on seed development in genetic line of peas, G_2 and its relation to changes in endogenous gibberellins measured by combined gas chromatography and mass spectrometry. Planta 146:423–432

Isikawa S (1954) Light sensitivity against the germination. I. "Photoperiodism" of seeds. Bot Mag 67:51–56

Isikawa S (1962) Light sensitivity against the germination. III. Studies on various partial processes in light sensitive seeds. Jpn J Bot 18:105–132

Jackson GAD (1968) Hormonal control of fruit development, seed dormancy and germination with particular reference to *Rosa*. In: Plant growth regulators. Soc Chem Ind Monogr 31:127–156

Jacobson R, Globerson D (1980) *Daucus carota* (carrot) seed quality, I Effects of seed size on germination, emergence and plant growth under subtropical conditions, II The importance of the primary umbel in carrot seed production. In: Hebblethwaite PD (ed) Seed production. Butterworth, London

Jacques R (1957) Quelques données sur le photoperiodisme des *Chenopodium polyspermum* L.; influence sur la germination des graines. Colloque International sur le Photo-thermo periodisme, Publ 34 Serie B. IUBS, Parma, pp 125–130

Jacques R (1968) Action de la lumière par l'intermediaire du phytochrome sur la germination, la croissance et le développement de *Chenopodium polyspermum* L. Physiol Vég 6:137–164

Johnson LPV (1935) General preliminary studies on the physiology of delayed germination in *Avena fatua*. Can J Res C 13:283–300

Juhren M, Went FW, Phillips E (1956) Ecology of desert plants, IV. Combined field and laboratory work on germination of annuals in the Joshua Tree National Monument, California. Ecology 37:318–330

Juntila O (1973) Seed and embryo germination in *Syringa vulgaris* and *S. reflexa* as affected by temperature during seed development. Physiol Plant 29:264–268

Kadman A (1954) Germination of some summer-annuals. M Sc Thesis, Hebrew University of Jerusalem (in Hebrew)

Kadman-Zahavi A (1960) Effects of short and continuous illuminations on the germination of *Amaranthus retroflexus* seeds. Bull Res Counc Isr, Sec D 9:1–20

Kamenetsky I (1987) Wild *Allium* in South Kazakhstan. Proc Kaz SSR Acad Sci, Nauka, Alma-Ata, Kazakhstan, pp 91–92

Kamenetsky I (1988) Biomorphological types and root systems of wild Allium of Kazakhstan. Problems of ecological morphology of plants. Nauka, Alma-Ata, Kazakhstan, pp 111–126

Kamenetsky R, Gutterman Y: Life cycles and delay of seed dispersal in some geophytes inhabiting the Negev desert highlands of Israel. J Arid Environ (in press)

Karssen CM (1970) The light promoted germination of the seeds of *Chenopodium album* L., III Effect of the photoperiod during growth and development of the plants on the dormancy of the produced seeds. Acta Bot Neerl 19:81–94

Karssen CM (1980/81) Environmental conditions and endogenous mechanisms involved in secondary dormancy of seeds. Isr J Bot 19:45–64

Karssen CM, Haigh A, van der Toorn P, Weges R (1989) Physiological mechanisms involved in seed priming. In: Taylorson RB (ed) Recent advances in the development and germination of seeds. Plenum, New York, pp 269–280

240 References

Kassas M (1966) Plant life in deserts. In: Hills ES (ed) Arid lands. Methuen, London, pp 145–180

Kay BL, Graves WL, Young JA (1988) Long-term storage of desert shrub seed. Mojave Reveg Notes 23:1–22

Keeley JE (1986) Seed germination patterns of *Salvia mellifera* in fire-prone environments. Oecologia 71:1–5

Keeley JE (1987) Role of fire on seed germination of woody taxa in California chaparral. Ecology 68:434–443

Keeley JE, Morton BA, Petrosa A, Trotter P (1985) Role of allelopathy, head and charred wood in the germination of chaparral herbs and suffrutescents. J Ecol 73:445–458

Keeley JE, Zedler PH, Zammit CA, Stohlgren TJ (1989) Fire and demography. The California chaparral. Paradigms reexamined. Nat Hist Mus Los Ang Cty Sci Ser 34:151–153

Kelly KM, van Staden J (1987) The lens as the site of permeability in the papilionoid seed, *Aspalathus linearis*. J Plant Physiol 128:395–404

Keren A, Evenari M (1974) Some ecological aspects of distribution and germination of *Pancratium maritimum* L. Isr J Bot 23:202–215

Kerley GIH (1991) Seed removal by rodents, birds and ants in the semi-arid Karoo, South Africa. J Arid Environ 20:63–69

Kermode AR, Bewley JD, Dasgupta J, Misra S (1986) The transition from seed development to germination: a key role for desiccation? HortScience 21:1113–1118

Kermode AR, Oishi MY, Bewley JD (1989a) Regulatory roles for desiccation and abscisic acid in seed development: a comparison of the evidence from whole seeds and isolated embryos. In: Seed moisture. Crop Sci Soc Am Spec Publ 14:23–50

Kermode AR, Pramanik K, Bewley JD (1989b) The role of maturation drying in the transition from seed development to germination. VI. Desiccation-induced changes in messenger RNA populations within the endosperm of *Ricinus communis* L. seeds. J Exp Bot 40:210, 33–41

Khan MA, Ungar IA (1984a) The effect of salinity and temperature on the germination of polymorphic seeds and growth of *Atriplex triangularis* Willd. Am J Bot 71:481–489

Khan MA, Ungar IA (1984b) Seed polymorphism and germination responses to salinity stress in *Atriplex triangularis* Willd. Bot Gaz 145:487–494

Khan MA, Ungar IA (1985) The role of hormones in regulating the germination of polymorphic seeds and early seedling growth of *Atriplex triangularis* under saline conditions. Physiol Plant 63:109–113

Kigel J (1992) Diaspore heteromorphism and germination in populations of the ephemeral *Hedypnosis rhagadioloides* (L.) F.W. Schmidt (Asteraceae) inhabiting a geographic range of increasing aridity. Acta Oecol 13:45–53

Kigel J, Ofi M, Koller D (1977) Control of the germination responses of *Amaranthus retroflexus* L. seeds by their parental photothermal environment. J Exp Bot 28:1125

Kigel J, Gibly A, Negbi M (1979) Seed germination in *Amaranthus retroflexus* L. as affected by the photoperiod and age during flower induction of the parent plants. J Exp Bot 30:997–1002

Kincaid RR (1935) The effects of certain environmental factors on germination of Florida cigar wrapper tobacco seeds. Fla Agr Exp Sta Bull 277:47

King TJ (1975) Inhibition of seed germination under leaf canopies in *Renaria sertyllifolia, Veronica arvensis* and *Cerastium holosteoides*. New Phytol 75:87

Kivilaan A, Bandurski RS (1981) The one hundred-year period for Dr. Beal's seed viability experiment. Am J Bot 68:1290–1292

Kolattukudy PE (1980a) Cutin, suberin, and waxes. In: Stumpf PK (ed) The biochemistry of plants, vol 4. Lipids: structure and function. Academic Press, New York

Kolattukudy PE (1980b) Biopolyster membranes of plants: Cutin and suberin. Science 208:990

Kolattukudy PE (1981) Structure, biosynthesis, and biodegradation of cutin and suberin. In: Briggs WR, Green PB, Jones RL (eds) Annu Rev Plant Physiol, vol 32. Annual Reviews, Palo Alto

Kolattukudy PE (1984) Biochemistry and function of cutin and suberin. Can J Bot 62:2918

Koller D (1954) Germination regulating mechanisms in desert seeds. Ph D Thesis, Hebrew University of Jerusalem (in Hebrew with English summary)

References

Koller D (1956) Germination regulating mechanisms in some desert seeds – III. *Calligonum comosum* L'her. Ecology 37:430–433

Koller D (1957) Germination-regulating mechanisms in some desert seeds. IV. *Atriplex dimorphostegia* Kar. et Kir. Ecology 38:1–13

Koller D (1962) Preconditioning of germination in lettuce at time of fruit ripening. Am J Bot 49:841–844

Koller D (1969) The physiology of dormancy and survival of plants in desert environments. Symp Soc Exp Biol 23:449–469

Koller D (1972) Environmental control of seed germination. In: Kozlowski TT (ed) Seed biology, II. Academic Press, London, pp 2–102

Koller D, Negbi M (1959) The regulation of germination in *Oryzopsis miliacea*. Ecology 40:20–36

Koller D, Negbi M (1966) Germination of seeds of desert plants. Final report to the U.S.D.A., pp 1–180

Koller D, Roth N (1964) Studies on the ecological and physiological significance of amphicary in *Gymnarrhena micrantha* (Compositae). Am J Bot 51:26–35

Koller D, Sachs M, Negbi M (1964) Germination – Regulating mechanisms in some desert seeds, VII. *Artemisia monosperma*. Plant Cell Physiol 5:85–100

Köppen W (1954) Classification of climates and the world patterns. In: Trewartha GT (ed) An introduction to climate, 3rd edn. McGraw-Hill, New York, pp 225–226; 381–383

Kurtz EB (1958) Chemical basis of adaptation in plants. Science 128:1115–1117

Lalonde L, Bewley JD (1986) Patterns of protein synthesis during the germination of pea axes, and the effects of an interrupting desiccation period. Planta 167:504–510

Lang A (1965) Effects of some internal and external conditions on seed germination. In: Ruhland W (ed) Encyclopedia of plant physiology, XV. Differentiation and development, part 2. Springer, Berlin Heidelberg New York, pp 893–894

Lavie D, Levy EC, Cohen A, Evenari M, Gutterman Y (1974) New germination inhibitor from *Aegilops ovata* L. Nature 249:388

le Houérou HN (1986) The desert and arid zones of northern Africa. In: Evenari M, Noy-Meir I, Goodall DW (eds) Ecosystems of the world, 12 B. Hot deserts and arid shrubland B. Elsevier, Amsterdam, pp 101–148

Leigh JH, Noble JC (1972) Riverine of New South Wales – its pastoral and irrigation development. Div Pl Ind, CSIRO Aust, Canberra

Lerner HR, Mayer AM, Evenari M (1959) The nature of the germination inhibitors present in dispersal units of *Zygophyllum* and *Trigonella arabica*. Physiol Plant 12:245–250

Liou TS (1987) Studies on germination and vigour of cabbage seeds. Ph D Thesis, Agricultural University, Wageningen, The Netherlands

Lona F (1947) L'influenza della condizioni ambientali durante l'embriogenesi sulla caratteristiche del seme e della pianta che ne deriva. Lavori di Botanica, vol pubbl in occasione del 70° genetliaco del Prof. Gola, pp 313–352

Loria M, Noy-Meir I (1979/80) Dynamics of some annual populations in a desert loess plain. Isr J Bot 28:211–225

MacMahon JA, Wagner FH (1985) The Mojave, Sonoran and Chihuahuan deserts of North America. In: Evenari M, Noy-Meir I, Goodall DW (eds) Hot deserts and arid shrublands. Elsevier, Amsterdam, pp 105–202

Mahmoud A, El-Sheikh AM, Abdul Baset S (1981) Germination of *Verbesina enceliodes* and *Rumex nervosus* from south Hijaz. J Arid Environ 4:299–308

Manning JC, van Staden J (1987) The role of the lens in seed imbibition and seedling vigour of *Sesbania punicea* (Cav.) Benth. (Leguminosae: Papilionoideae). Ann Bot 59:705–713

Marchaim U, Werker E, Thomas WDE (1974) Changes in the anatomy of cotton seed coats caused by lucerne saponins. Bot Gaz 135:139–146

Mares MA, Rosenzweig ML (1978) Granivory in North and South American deserts: rodents, birds and ants. Ecology 59:235–241

Mayer AM, Poljakoff-Mayber A (1982) The germination of seeds, 3rd edn. Pergamon, Oxford

McCullough JM, Shropshire W Jr (1970) Physiological predetermination of germination responses in *Arabidopsis thaliana* (L.) Heynh. Plant Cell Physiol 11:139–148

Meneghini M, Vicente M, Noronha AB (1968) Effect of temperature on dark germination of *Rumex obtusifolius* L. seeds. A tentative physico-chemical model. Arq Inst Biol Sao Paulo 35:33

Molisch H (1937) Über den Einfluß des Tabakrauchs auf die Pflanze. S.-B. Akad Wiss Wien Math-Nat KI 120:813−838

Monod Th (1986) The Sahel zone north of the equator. In: Evenari M, Noy-Meir I, Goodall DW (eds) Ecosystems of the world, 12 B. Hot deserts and arid shrublands B. Elsevier, Amsterdam, pp 203−244

Morton SR (1985) Granivory in arid regions: Comparison of Australia with North and South America. Ecology 66:1859−1866

Motro U (1982) Optimal rates of dispersal, 1. Haploid populations, 2. Diploid populations. Theor Popul Biol 21:394−411, 412−429

Mott JJ, Groves RH (1981) Germination strategies. In: Pate JA, McComb AJ (eds) Biology of Australian plants. University of Western Australia Press, Australia, pp 307−341

Müller-Schneider P (1967) Zur Verbreitungsbiologie des Moschuskrautes (*Adoxa moschatellina*). Vegetatio 15:27−32

Murbeck S (1919/20) Beiträge zur Biologie der Wüstenpflanzen, I, II. Lunds Univ. Arsskr. 15,10 and 17,1

Nagao M, Esashi Y, Tanaka T, Kumagai T, Fukumoto S (1959) Effects of photoperiod and gibberellin on germination of seeds of *Begonia evansiana* Andr. Plant Cell Physiol 1:39

Naveh Z (1973) The ecology of fire in Israel. In: Rommarek R (ed) Proc 13th Tall Timbers Fire Ecology Conf, Tallahasee, pp 137−170

Naveh Z (1974) Effects of fire in the Mediterranean region. Fire in ecosystems. Academic Press, New York, pp 401−434

Nevo E, Beiles A, Gutterman Y, Storch N, Kaplan D (1984a) Genetic resources of wild cereals in Israel and the vicinity: II Phenotypic variation within and between populations of wild wheat, *Hordeum spontaneum*. Euphytica 33:737−756

Nevo E, Beiles A, Gutterman Y, Storch N, Kaplan D (1984b) Genetic resources of wild cereals in Israel and the vicinity: I Phenotypic variation within and between populations of wild wheat, *Iriticune dicoccoides*. Euphytica 33:717−735

Nikolaeva MG (1969) Physiology of deep dormancy in seeds. Israel Program of Scientific Translations, Jerusalem (Available from US Dept. of Commerce)

Noble JC (1975a) Differences in size of emus on two contrasting diets on the Riverine Plain of New South Wales. Emu 75:35−37

Noble JC (1975b) The effects of emus (*Dromaius novaehollandiae* Latham) on the distribution of the nitre bush (*Nitraria billardieri* DC.). J Ecol 63:979−984

Nordhagen R (1936) Über dorsiventrale und transversale Tangentballisten. Sven Bot Tidskr 30:443−473

O'Dowd DJ, Hay ME (1980) Mutualism between harvester ants and a desert ephemeral: seed escape from rodents. Ecology 61:531−540

Ofer Y (1982) The influence of ants on the plant composition in grazing area. ROTEM − Bulletin of the Israel Plant Information Center 3, Society for the Protection of Nature in Israel, Tel-Aviv, pp 48−51 (in Hebrew with English summary, pp 79−80)

Orshan G (1986) The deserts of the Middle East. In: Evenari M, Noy-Meir I, Goodall DW (eds) Ecosystems of the world, 12 B. Hot deserts and arid shrublands B. Elsevier, Amsterdam, pp 1−28

Orshan G (1989) Description of plant annual cycles. In: Orshan G (ed) Plant phenomorphological studies in Mediterranean type ecosystems (Geobotany 12). Kluwer Academic, London, pp 99−157

Overbeck F (1925) Über den Mechanismus der Samenabschleuderung von *Cardamine impatiense*. Ber Dtsch Bot Ges 43:469

Owen JG (1987) On productivity as a predator of rodent and carnivore diversity. Ecology 69:1161−1165

Philipupillai J, Ungar IA (1984) The effect of seed dimorphism on the germination and survival of *Salicornia europaea* L. populations. Am J Bot 71:542−549

References

Porsild AE, Harington CR, Mulligan GA (1967) *Lupinus arcticus* Wats, grown from seeds of Pleistocene age. Science 158:113–114

Pourrat Y, Jacques R (1975) The influence of photoperiodic conditions received by the mother plant on morphological and physiological characteristics of *Chenopodium polyspermum* L. seeds. Plant Sci Lett 4:273–279

Quail PH, Carter OG (1969) Dormancy of seeds of *Avena ludoviciana* and *A. fatua*. Aust J Agric Res 20:1–11

Quinlivan BJ (1961) The effect of constant and fluctuating temperatures on the permeability of the hard seeds of some legume species. Aust J Agric Res 12:1009–1022

Quinlivan BJ (1965) The influence of the growing season and the following dry season on the hard-seededness of subterranean clover in different environments. Aust J Agric Res 16:277–291

Quinlivan BJ (1966) The relationship between temperature fluctuations and the softening of hard seeds of some legume species. Aust J Agric Res 17:625–631

Quinlivan BJ (1968) The softening of hard seeds of sand-plain lupin (*Lupinus varius* L.). Aust J Agric Res 19:507–515

Ramakrishnan PS, Kapoor P (1974) Photoperiodic requirements of seasonal populations of *Chenopodium album* L. J Ecol 62:67–73

Rao NK, Roberts EH, Ellis RH (1987) The influence of pre- and poststorage hydration treatments on chromosomal aberrations, seedling abnormalities, and viability of lettuce seeds. Ann Bot 60:97

Reichman OJ (1975) Relationship of desert rodent diets to available resources. J Mammal 56:731–751

Reichman OJ (1979) Desert granivore foraging and its impact on seed densities and distributions. Ecology 60:1085–1092

Reichman OJ (1984) Spatial and temporal variations of seed distributions in Sonoran Desert soils. J Biogeo 11:1–11

Richter R, Libbert E (1967) Investigations on the germination-behaviour of some halophytic plants occurring at the coat of the Baltic Sea. In: Borris H (ed) Physiologie, Oekologie und Biochemie der Keimung 1. Ernst-Moritz-Arndt-Universität, Greifswald, pp 459–469

Rissing SW (1986) Indirect effects of granivory by harvester ants: plant species composition and reproductive increase near ant nests. Oecologia 68:231–234

Roach DA, Wulff RD (1987) Maternal effects in plants. Annu Rev Ecol Syst 18:209–235

Roberts HA, Feast PM (1973) Emergence and longevity of seeds of annual weeds in cultivated and undisturbed soil. J Appl Ecol 10:133–143

Roberts HA, Neilson JE (1982) Seasonal changes in the temperature requirements for germination of buried seeds of *Aphanes arvensis* L. New Phytol 92:159–166

Roiz L (1989) Sexual Strategies in some gynodioecious and gynomonoecious plants. Ph D Thesis, Tel-Aviv University, Tel-Aviv

Rollin P (1972) Phytochrome control of seed germination. In: Mitrakos K, Shropshire W (eds) Phytochrome. Academic Press, London, pp 228–254

Rolston MP (1978) Water impermeable seed dormancy. Bot Rev 44:365

Roth-Bejerano N, Koller D, Negbi M (1971) Photocontrol of germination in *Hyoscyamus desertorum* a kinetic analysis. Isr J Bot 20:28–40

Rudloff W (1981) World climates. Wissenschaftliche Verlagsgesellschaft, Stuttgart

Rutherford MC, Westfall RH (1986) The biomes of southern Africa – an objective categorization. Mem Bot Surv S Afr 54:1–98

Sabo DG, Johnson GV, Martin WC, Aldon EF (1979) Germination requirements of 19 species of arid land plants. Rocky Mountain Forest and Range Exp Stn, Forest Service, US Dept Agriculture. SEAM (Surface Environment and Mining) USDA Forest Service Program, Res Pap RM-210:1–26

Saini HS, Consolocion ED, Bassi PK, Spencer MS (1986) Requirement for ethylene synthesis and action during relief of thermoinhibition of lettuce seed germination by combinations of gibberellic acid, kinetin, and Carbon Dioxide. Plant Physiol 81:950–953

Shachak M (1975) Some aspects of the structure and function of a desert ecosystem and its use in a teaching programme of a field studies center. Ph. D. Thesis, Hebrew University of Jerusalem (in Hebrew with English summary)

Shachak M, Brand S (1988) Relationship among settling, demography and habitat selection: an approach and a case study. Oecologia 76:620–627

Shachak M, Brand S (1991) Relations among spatiotemporal heterogeneity, population abundance, and variability in a desert. In: Kolasa J, Pickett S (eds) Ecological heterogeneity. Springer, Berlin Heidelberg New York, pp 202–223

Shachak M, Brand S, Gutterman Y (1991) Porcupine disturbances and vegetation pattern along a resource gradient in a desert. Oecologia 88:141–147

Shmida A (1984) Why do some Compositae have a deciduous pappus? Ann Mo Bot Garden 72:184–186

Shmida A, Evenari M, Noy-Meir I (1986) Hot desert ecosystems: an integrated view. In: Evenari M, Noy-Meir I, Goodall DW (eds) Ecosystems of the world, 12B. Hot deserts and arid shrublands B. Elsevier, Amsterdam, pp 379–388

Silvertown JW (1982) Introduction to plant population ecology. Longman, London

Small JGC, Gutterman Y (1991) Evidence for inhibitor involvement in thermodormancy of Grand Rapids lettuce seeds. Seed Sci Res 1:263–267

Small JGC, Gutterman Y (1992a) Effect of sodium chloride on prevention of thermodormancy and effects on ethylene and protein synthesis and respiration in Grand Rapids lettuce seeds. Physiol Plant 84:35–40

Small JGC, Gutterman Y (1992b) A comparison of thermo- and skotodormancy in seeds of *Lactuca serriola* in terms of induction, alleviation, respiration, ethylene and protein synthesis. Plant Growth Regulation (in press)

Stebbins KL (1977) Flowering Plants Evolution Above the Species Level. Belkman/Harvard University Press, Cambridge, Mass

Steinbrinck C, Schinz H (1908) Über die anatomische Ursache der hygrochastischen Bewegungen der sog. Jerichorosen USW. Flora 98:471–500

Stokes P (1965) Temperature and seed dormancy. In: Ruhland W (ed) Encyclopedia of plant physiology, XV. Differentiation and development, part 2. Springer, Berlin Göttingen Heidelberg, pp 746–803

Styer RC, Cantliffe DJ, Hall CB (1980) The relationship of ATP concentration to germination and seedling vigor of vegetable seeds stored under various conditions. J Am Hortic Sci 105:298–303

Takahashi K, Arakawa H (eds) (1981) Climates of southern and western Asia, world survey of climatology, vol 9. Elsevier, Amsterdam, pp 1–333

Taylorson RB (1972) Phytochrome controlled changes in dormancy and germination of buried weed seeds. Weed Sci 20:417–422

Tevis L (1958a) Interrelations between the harvester ant *Veromessor pegandei* (Mayr) and some desert ephemerals. Ecology 39:695–704

Tevis L (1958b) A population of desert ephemerals germinated by less than one inch of rain. Ecology 39:688–695

Tevis L (1958c) Germination and growth of ephemerals induced by sprinkling a sandy desert. Ecology 39:681–688

Thomas TH, Biddington NL, O'Toole DF (1979) Relationship between position on the parent plant and dormancy characteristics of seeds of three cultivars of celery (*Apium graveolens*). Physiol Plant 45:492–496

Thompson PA (1973a) Geographical adaptation of seeds. In: Heydecker W (ed) Seed ecology. Butterworth, London, pp 31–58

Thompson PA (1973b) Seed germination in relation to ecological and geographical distribution. In: Heywood VH (ed) Taxonomy and ecology, Spec Vol V. Syst Assoc, Academic Press, London, pp 93–119

Thurling N (1966) Population differentiation in Australian cardamine. Aust J Bot 14:189–194

Thurston JM (1951) Biology of wild oats. Rep Rothamsted Exp Stn:67–69

Thurston JM (1962) An international experiment on the effect of age and storage conditions on viability and dormancy of *Avena fatua* seeds. Weed Res 2:122–129

Tilman D (1982) Resource competition and community structure. Princetown University Press, Princeton

References

Toole VK (1973) Effects of light, temperature and their interactions on the germination of seeds. Seed Sci Technol 1:339–396

Trewartha GT (1961) The earth's problem climates. University of Wisconsin Press, Wisconsin, pp 1–334

Turesson G (1922) The genotypical responses of the plant species to habitat. Hereditas 3:211–350

Ulbrich E (1928) Biologie der Früchte und Samen (Karpobiologie). Springer, Berlin

Ungar IA (1978) Halophyte seed germination. Bot Rev 44:233–264

Ungar IA (1979) Seed dimorphism in *Salicornia europaea* L. Bot Gaz 140:102–108

Van der Pijl L (1982) Principles of dispersal in higher plants, 3rd edn. Springer, Berlin Heidelberg New York, pp 1–24

Vander Veen R (1970) The importance of the red-far red antagonism in photoblastic seeds. Acta Bot Neerl 19:809

Van der Woude W (1989) Phytochrome and sensitization in germination control. In: Taylorson RB (ed) Recent advances in the development and germination of seeds. Plenum, New York, pp 181–189

Van de Venter HA, Esterhuizen AD (1988) The effect of factors associated with fire on seed germination of *Erica sessiliflora* and *E. hebecalyx* (Ericaceae). S Afr J Bot 54:301–304

Van Rooyen MW, Theron GK, Grobbelaar N (1990) Life form and dispersal spectra of the flora of Namaqualand, South Africa. J Arid Environ 19:133–145

Venable DL (1985) The evolutionary ecology of seed heteromorphism. Am Nat 126:577–595

Venable DL, Lawlor L (1980) Delayed germination and dispersal in desert annuals: escape in space and time. Oecologia 46:272–282

Venable DL, Levin DA (1985) Ecology of achene dimorphism in *Heterotheca latifolia*: achene structure, germination and dispersal. J Ecology 73:133–145

Venable DL, Búrquez A, Corral G, Morales E, Espinosa F (1987) The ecology of seed heteromorphism in *Heterosperma pinnatum* in central Mexico. Ecology 68:65–76

Vickery RK (1967) Ranges of temperature tolerance for germination of *Mimulus* seeds from diverse populations. Ecology 48:647–651

Vidaver W, Hsiao AI (1975) Secondary dormancy in light-sensitive lettuce seeds incubated anaerobically or at elevated temperature. Can J Bot 53:2557–2560

Von Guttenberg H (1926) Die Bewegungsgewebe. In: Linsbauer K (ed) Handbuch der Pflanzenanatomie, vol 5, Sect 1 Part 2. Borntraeger, Berlin

Wagenitz G (1969) Abgrenzung und Gliederung der Gattung *Filago* L. s.l. (Compositae-Inulae). Willdenowia 5:395–444

Wallace A, Rhoads WA, Frolich EF (1968) Behavior of *Salsola* as influenced by temperature, moisture, depth of planting and gamma irradiation. Agron J 60:76–78

Walter H (1986) The Namib Desert. In: Evenari M, Noy-Meir I, Goodall DW (eds) Ecosystems of the world, 12B. Hot deserts and arid shrublands B. Elsevier, Amsterdam, pp 245–282

Warburg D, Eig A (1926) *Pissum fulvum* Sibth. et Smith n. var amphicarpum. Agric Rec Tel-Aviv 1:1–6

Warburg O (1892) Ueber Ameisenpflanzen (Myrmecophyten). Biol Zbirnik L'vivs'kii Derzhavnii Univ 12:129–142

Wareing PF (1965) Endogenous inhibitors in seed germination and dormancy. In: Ruhland W (ed) Encyclopedia of plant physiology, XV. Differentiation and development, part 2. Springer, Berlin Göttingen Heidelberg, pp 909–924

Went FW (1948) Ecology of desert plants, I. Observations on germination in the Joshua Tree National Monument, California. Ecology 29:242–253

Went FW (1949) Ecology of desert plants, II. The effect of rain and temperature on germination and growth. Ecology 300:1–13

Went FW (1953) The effects of rain and temperature on plant distribution in the desert. Proc Int Symp on Desert research, May 1952. Res Council Isr Spec Publ 2:230–240

Went FW (1957) Experimental control of plant growth. Chronica Botanica, Waltham, MA, pp 248–251

Went FW (1961) Problems in seed viability and germination. Proc Int Seed Test Ass 26:674–685

Went FW (1969) A long term test of seed longevity II. Aliso 7:1–12

Went FW, Munz PA (1949) A long term test of seed longevity. Aliso 2:63–75

246 References

Went FW (1969) A long term test of seed longevity II. Aliso 7:1–12

Went FW, Munz PA (1949) A long term test of seed longevity. Aliso 2:63–75

Went FW, Westergaard M (1949) Ecology of desert plants. III. Development of plants in the Death Valley National Monument, California. Ecology 30:26–38

Wentland MJ (1965) The effect of photoperiod on the seed dormancy of *Chenopodium album*. PhD Thesis, University of Wisconsin, Madison

Werger MJA (1986) The Karoo and Southern Kalahari. In: Evenari M, Noy-Meir I, Goodall DW (eds) Ecosystems of the world, 12B. Hot deserts and arid shrublands B. Elsevier, Amsterdam, pp 283–360

Werger MJA, Coetzee BJ (1978) The Sudano-Zambezian region. In: Werger MJA (ed) Biogeography and ecology of southern Africa. W Junk, The Hague, pp 301–462

Werker E (1980/81) Seed dormancy as explained by the anatomy of embryo envelopes. Isr J Bot 29:22

West ES (1952) A study of the annual soil temperature wave. Aust J Sci Res A5:303–314

Whitford WG (1978) Structure and seasonal activity of Chihuahuan desert ant communities. Insectes Soc 25:79–88

Whittaker RH (1956) Vegetation of the Great Smoky Mountains. Ecol Monogr 26:1–80

Whittaker RH (1967) Gradient analysis of vegetation. Biol Rev 42:207–264

Wicklow DT (1977) Germination response in *Emmenanthe penduliflora* (Hydrophyllaceae). Ecology 58:201–205

Witztum A, Gutterman Y, Evenari M (1969) Integumentary mucilage as an oxygen barrier during germination of *Blepharis persica* (Burm.) Kuntze. Bot Gaz 130:238–241

Woolley JT, Stoller EW (1978) Light penetration and light-induced seed germination in soil. Plant Physiol 61:597–600

Wurzburger J, Koller D (1976) Differential effects of the parental photothermal environment on development of dormancy in caryopses of *Aegilops lotschyi*. J Exp Bot 27:43–48

Yair A, Shachak M (1987) Studies in watershed ecology of an arid area. In: Berkofsky L, Wurtele MG (eds) Progress in desert research. Rowman and Littlefield, Totowa, New Jersey, pp 145–193

Zangvil A, Druian P (1983) Meteorological data for Sede Boqer. The Jacob Blaustein Institute for Desert Research, Ben-Gurion University of the Negev Israel. Desert Meteorology Pap, Ser A, No 8

Zeevart JAD (1966) Reduction of the gibberellin content of pharbitis seeds by CCC and after-effects in the progeny. Plant Physiol 41:856

Zohar Y, Waisel Y, Karshon Y (1975) Effect of light, temperature and osmotic stress on seed germination of *Eucalyptus occidentalis* Endl. Aust J Bot 23:391–397

Zohary D, Hopf M (1988) Domestication of plants in the Old World. Oxford Science Publications; Clarendon, Oxford

Zohary M (1937) Die verbreitungsökologischen Verhältnisse der Pflanzen Palästinas. Die antitelechorischen Erscheinungen. Beih Bot Zentrale 56:1–55

Zohary M (1962) Plant life of Palestine. Ronald, New York

Zohary M (1966) Flora Palestina, Part One-Text. The Israel Academy of Sciences and Humanities, Jerusalem

Zohary M (1972) Flora Palestina, Part Two-Text. The Israel Academy of Sciences and Humanities, Jerusalem

Zohary M, Fahn A (1941) Anatomical-carpological observations in some hygrochastic plants of the oriental flora. Palest J Bot 2:125–135

Latin Name Index

PLANTS

Acacia 99, 228
- *gerrardii* 99
- *raddiana* 99, 152
- *tortilis* 99
Acanthus syriacus 133
Aegilops 141
- *geniculata* 31, 32*, 32, 33, 34, 35, 41, 70, 74, 76, 80*, 90, 92*, 92, 148*, 180
- *kotschyi* 42
- *ovata* (see *A. geniculata*)
Aellenia autrani (see *Halothamnus hierochunticus*)
Agave lecheguilla 197
Agropyron smithii 197
Aizoon 141, 227
- *canariense* 105, 106, 111, 113
- *hispanicum* 103, 105, 106, 109, 111, 113, 205
Allium spp. 19, 148*
- *rothii* 90
- *schuberti* 80*, 90
Alyssum damascenum 106, 110
Amaranthus fimbriatus 177, 191
- *retroflexus* 41, 200, 204
Ammi visnaga 108, 112
Anastatica hierochuntica 103, 106, 110, 110*, 121, 141, 148*, 188, 220, 221
Andropogon scoparius 197
Anemone coronaria 86
Anthemis pseudocotula 106, 110, 114
Antirrhinum coulterianum 182
Anvillea garcinii 106, 111, 114
Aphanes arvensis 164
Arabidopsis kneuckeri 84
- *pumila* 84
- *thaliana* 62, 199
Artemisia frigida 198
- *herba-alba* (see *A. sieberi*)
- *monosperma* 88, 138, 150, 189, 200, 201, 203, 204

- *sieberi* (= *A. herba-alba*) 15, 20, 91, 102, 138, 140, 143, 148*, 150, 188, 189, 201, 211, 216, 219, 221, 222*, 227, 228
- *tridentata* 198
Aspalathus linearis 157
Aster tripolium 202
Asteriscus aquaticus 108, 111
- *graveolens* 106, 110, 114
- *pygmaeus* 18, 36, 76, 106, 111, 115, 115*, 116, 117*, 118*, 119*, 120*, 121*, 122*, 141, 147*, 164, 166, 170, 171, 172*, 229
Astragalus asterias (= *A. cruciatus*) 106, 111
- *cruciatus* (see *A. asterias*)
- *hispidulus* 209
- *sinicus* 41
- *spinosus* 80*, 85, 89
- *tribuloides* 103, 106, 111, 209
Atriplex confertifolia 198
- *dimorphostegia* 39, 150, 160, 180, 200, 204
- *halimus* 180, 181
- *holoscarpa* (= *A. spongiosa*) 39
- *inflata* 39
- *litoralis* 202
- *rosea* 39
- *semibaccata* 39
- *spongiosa* (see *A. holoscarpa*)
- *triangularis* 74
- *vesicaria* 99
Avena barbata 92
- *fatua* 160
- *ludoviciana* 160
- *sterilis* 92
- *wiestii* 92
Avicennia marina 165

Bellevalia desertorum 11, 18, 84, 85, 97, 147*, 194, 195, 196*
- *eigii* 11, 18, 80*, 84, 86*, 98, 147*, 194, 196*

---

* Figures

248 Latin Name Index

Bergeranthus 193
– *scapiger* 179*, 192, 200
Beta vulgaris 96
Betula pubescens 201
Bidens pilosa 70
Blepharis attenuata 107, 125, 128, 134, 136
– *ciliaris* 107, 123*, 128, 134, 136, 137
– *linearifolia* 107, 123*
– *spp.* 18, 21, 76, 103, 107, 112, 122, 123*, 124*, 125, 126*, 127*, 129*, 130*, 130, 131*, 132*, 133*, 133, 134, 135*, 136, 136*, 137, 140, 141, 143, 148*, 170, 171, 172*, 188, 189, 201, 203, 216, 227, 228, 230
Boerhavia spicata 177
Bouteloua aristidoides 191
– *barbata* 191
– *curtipendula* 197
– *gracilis* 197
Bromus fasciculata 80*, 89
– *rubens* 191, 219, 220*, 220

Cakile maritima 202
Calendula arvensis 80*, 89*, 89
Calligonum comosum 150, 198, 202
Calotropis procera 86, 87*
Calyptridium monandium 177
Caralluma europaea (= *C. negevensis*) 86
– *negevensis* (see *C. europaea*)
Cardamine 22
Carex pachystylis 11, 18, 19*
Carnegiea gigantea 160, 197, 200
Carrichtera annua 50, 51*, 51, 53, 57, 64, 76, 80, 82*, 83*, 106, 110, 122, 141, 148*, 150, 173*, 185, 186, 187*, 194, 201, 209, 212*, 213
Cercidium aculeatum 152, 169
Cercocarpus montanus 198
Chaenactis artemisiaefolia 182
– *caryphoclinia* 177
– *fremontii* 177
Cheiridopsis aurea 64*, 64, 65*
– *spp.* 64, 66*, 77, 113, 178*, 192
Chenopodium album 22, 42, 53, 57, 66, 191, 200, 204
– *bonus-henricus* 53, 61
– *polyspermum* 52, 53, 57
Chrysanthemum coronarium 96
Chrysothamnus nauseosus ssp.
– *bigelovii* 198
– *nauseosus* spp. *consimilis* 198
Cichorium pumilum 108, 111
Citrullus colocynthis 80*, 85
Colchicum ritchii 90
– *tunicatum* 20, 80*, 90

Coreopsis bigelovii 160, 177
Cowania stansburiana 198
Crepis 87
Cumumis prophetarum 54, 55*, 65, 66, 68*, 68, 69*, 148*, 199, 200
– *sativus* 55, 56, 65, 66, 68, 69*, 70*, 75, 76, 199
Cutandia memphitica 104
Cyperus inflexus 22, 164

Dalea spinosa 152, 169
Dimorphotheca polyptera 39, 139
– *sinuata* 139
Diplotaxis acris 84
– *harra* 84, 210, 211, 215, 216
Distichlis stricta 197

Echium judaeum 96
Emex spinosa 27, 27*, 28, 76, 138, 139, 143, 144, 145, 171, 172*, 180, 195, 221
Emmenanthe penduliflora 182
Erica hebecalyx 182
– *sessiliflora* 182
Eriophyllum confertiflorum 182
– *wallacei* 160, 191
Erodium bryoniifolium (see *E. oxyrhynchum*)
– *crassifolium* (= *E. hirtum*) 19*, 19, 80*, 93, 93*, 104
– *hirtum* (see *E. crassifolium*)
– *oxyrhynchum* (= *E. bryoniifolium*) 93, 175, 209
Erucaria boveana (see *E. rostrata*) 106, 110
– *rostrata* (= *E. boveana*)
Euphorbia micromera 191
– *polycarpa* 160

Fallugia paradoxa 198
Filago contracta 106, 111, 114
– *desertorum* 79, 82*, 87, 104, 144, 207, 209, 219*, 219, 220, 220*
Fouquieria splenden 193, 195, 197, 221

Geraea canescens 160
Geropogon 87
Gilia aurea 191
Glottiphyllum linguiforme 28, 29*, 30*, 113
Gomphocarpus sinaicus 86
Gundelia tournefortii 90
Gymnarrhena micrantha 26, 27*, 106, 111, 114, 114*, 138, 139, 141, 143, 145, 148*, 171, 172*, 200, 209, 216, 217, 221, 228
Halothamnus hierochunticus (= *Aellenia autrani*) 73, 74

* Figures

Latin Name Index

Hammada scoparia 80*, 90
Hedypnois cretica (= H. rhagadioloides) 39, 74
Hedypnois rhagadioloides (see H. cretica)
Helianthemum 21, 81, 140, 148*, 185
– kahiricum 96
– ventosum 21, 80*, 85, 95, 96, 183, 184, 184*, 185*, 185, 186, 190, 195
– vesicarium 21, 85, 183, 184, 184*, 185, 185*, 186*, 190, 195
Herniaria hirsuta 102, 104
Herrea elongata 94
Heterotheca latifolia 73, 87, 139
Hilaria Jamesii 197
Hirschfeldia incana 71, 72
Homeria schlechteri 94
Hordeum 141
– spontaneum 20, 80*, 90, 92, 148*, 158, 159*
Hyoscyamus desertorum 150, 200, 203

Ifloga rueppellii 87, 144
– spicata 87, 144
Ipomea stolonifera 202
Iris petrana 91*, 91, 100
Ixiolirion tataricum 90

Juttadinteria proximus 67*, 77

Lactuca sativa 33, 34, 54, 56, 57, 180
– serriola 25, 54, 57, 58*, 58, 59*, 60*, 61*, 62*, 63*, 71, 75, 77, 150, 191, 192, 201, 203
Lappula redowski 177
– spinocarpas 213
Larrea tridentata 221
Lathyrus aphaca 154
– hierosolymitanus 155
Lepidium aucheri 106, 110
– lasiocarpum 160, 177
– spinescens 108, 110
– spinosum 108, 110
Leptaleum filifolium 106, 111
Lophochloa berythea (see Rostraria smyrnaceae)
Loranthus acaciae 100, 165
Lupinus arboreus 153, 155*, 158
– arcticus 164
– pilosus (= L. varius)
– varius (see L. pilosus) 157
Lycopersicum esculentum 53, 54, 64
Malva aegyptia 80*, 84, 95, 176
– parviflora 96
Medicago laciniata 21, 148*, 154, 157

Menodora scabra 198
Mesembryanthemum 141, 227
– crystallinum 106, 111, 113
– forsskalii 105, 106, 111, 113
– nodiflorum 39, 40, 76, 105, 106, 111, 113, 140, 143, 148*, 161, 163*, 164, 166, 170, 172*, 173, 176, 176*, 180, 182, 183, 204, 205, 228
Mimulus 197
Mollugo cerviana 191
Muhlenbergia wrightii 197

Nasturtiopsis arabica (see N. coronopifolia)
– coronopifolia (= N. arabica) 84
Necacladus longiflorus 191
Nemophila insignis 190
Neotorularia torulosa (= Torularia torulosa) 106, 111
Nitraria billardieri 99, 152
– schoberi 99
Noea mucronata 102
Notoceras bicorne 106, 110

Oldenlandia corymbosa 41
Olneya tesota 152, 169
Ononis sicula 41, 42, 43, 44*, 45*, 45, 46, 46*, 47*, 50, 53, 56, 75, 80*, 84, 93, 95, 97, 151, 152, 154, 158
Ornithogalum trichophyllum 18, 19*
Ornithopus compressus 157
Oryzopsis miliacea (see Piptatherum miliaceum)

Pallenis spinosa 108, 110
Pancratium maritimum 88, 150, 198, 200, 201, 202
– sickenbergeri 80*, 89, 91
Panicum turgidum 160
Papaver humile 80*, 90
Parietaria diffusa (see P. judaica)
– judaica (= P. diffusa) 36, 37*, 38*
Pectis angustifolia 177
– papposa 191
Phagnalon rupestre 108, 110
Phaseolus vulgaris 73
Picris 87
– cyanocarpa 219*, 219, 220*, 220
Piptatherum miliaceum (= Oryzopsis miliaceae) 195, 200
Plantago bellardi 108, 112
– coronopus 95, 97, 97*, 98*, 99*, 103, 107, 112, 143, 148*, 200, 204, 205, 209, 219*

---

* Figures

250 Latin Name Index

- *crassifolia* 108, 112
- *cretica* 108, 112
- *insularis* 160
- *spinulosa* 191
Pleiospilos bulossii 113
Poa sinaica 19*
Polypogon monspeliensis 50
Portulaca oleracea 51, 52*, 57, 63, 67, 71, 75, 77, 150, 161, 162, 200, 201
Prosopis juliflora 95
Pteranthus dichotomus 29, 30*, 31, 74, 76, 80*, 89, 148*

Quercus calliprinos 165

Reaumuria hirtella 86
- *negevensis* 86
Reboudia pinnata 95, 96, 103, 106, 110, 141, 193, 193*, 194, 209, 213, 219*
Retama raetam 88, 99, 100*, 152
Ricinus communis 73
Roemeria hybrida 102, 104, 209
Rostraria smyrnaceae (= *Lophochloa berythea*) 219*
Rumex cyprius (= *R. roseus*) 180, 213
- *obtusifolium* 195
- *roseus* (see *R. cyprius*)

Salicornia europaea 37, 139
Salsola inermis 19, 19*, 21*, 90, 102, 181, 208, 209, 210, 212*, 213, 214, 215, 220
- *kali* 188
- *volkensii* 73, 90, 213
Salvia columbariae 22, 96, 160
- *horminum* 108, 112
- *lanigera* 95
- *viridis* 108, 112
Sarcobatus vermiculatus 198
Sarcopoterium spinosum 22
Scabiosa porphyroneura 80*, 88*, 88
Schismus arabicus 18, 79, 80*, 81*, 84, 96, 97, 104, 140, 143, 171, 172*, 175, 176, 209, 210, 212*, 219*, 230
Scilla hanburyi 19, 19*, 20, 89, 91
Scorzonera judaica 19, 21*, 86
- *papposa* 19, 80, 81, 83*, 84*, 86
Senecio 80*, 86
- *glaucus* 95
Sesbania panicea 158
Silybum marianum 96
Sisymbrium altissimum 160
Spegularia diandra 79, 81*, 84, 102, 104, 175, 209
Spheralcea incana 198

Spinifex hirsutus 202
Sporobolus contractus 197
- *cryptandrus* 197
Sternbergia clusiana 100, 101*
Stipa capensis 18, 80*, 92, 104, 109
Stipagrostis sabulicola 222
Streptanthus arizonicus 160

Tamarix 20, 21*, 80*, 86, 186
Thymelaea hirsuta 95
Torularia torulosa (see *Neotorularia torulosa*)
Trachomitum venetum 86
Trifolium pratense 153, 158
- *repens* 153, 153*, 156*, 158
- *subterraneum* 41, 60, 152, 157, 157*
Trianthema hereroensis 222
Trigonella arabica 21*, 41, 43, 46, 47*, 48*, 48, 49, 49*, 51, 53, 56, 75, 76, 89, 104, 112, 148*, 152, 154, 158, 181, 214*
- *stellata* 18, 103, 105, 106, 111, 112, 143, 205, 209
Triticum dicoccoides 21
Tulipa polychroma 90
- spp. 148*
- *systola* 18, 21*, 80*, 81, 85*, 90, 97, 98, 147*, 194, 195, 196*

Urginia maritima 20, 89, 91
- *undulata* 20, 21*, 89, 91
Ursinia cakilefolia 139

Verbascum blattaria 164
Verbesina enceliodes 22
Viscum cruciatum 100, 165, 228

Zygophyllum coccineum 202
- *dumosum* 20, 21*, 81, 85*, 90, 102, 140, 143, 180, 181, 189, 221, 227

ANIMALS AND INSECTS
(English names in Subject Index)

Capra ibex nubiana 99
Dromaius novaehollandiae 99
Gazella dorcas 99
Hystrix africaeaustralis 93
Hystrix indica 16, 93
Lepus capensis 99
Lepus europeus 99
Messor arenarius 95, 115
Messor eveninus 95, 96, 110, 115
Pogonomyrmex rugosus 96
Veromessor pergondei 96

* Figures

Subject Index

aerial 'seeds' 15, 26−28, 76, 114, 138−139, 171, 180, 195, 216, 221, 228
after-ripening 23, 40, 71
age of plants 25, 42, 151, 213, 214
age of seeds 50, 142
allelopathy 182
altitude(s) 25, 52−53, 61−62, 183
amphicarpy 26
animal(s) 27, 79−81, 85, 89, 96−97, 100, 103, 112, 114, 137, 147, 149, 152, 203, 207, 218
ants 15, 80−81, 84, 91, 95−97, 100−101, 110, 115, 120, 138−141, 149
aril(s) 91, 100−101
atelechoric 12, 15, 26, 28, 110−111, 113, 143, 205
atelechory 110, 138
autecological adaptation 17, 20
awn(s) 92, 159

ballist seed 90
balloon-like fruit 89
bi-seasonal annual 19, 215
bird 15, 81, 95−98, 100−101, 122, 138, 140, 142, 149, 165, 228
biotic factors 14

camel(s) 99, 136
capsule 28−30, 39−40, 73, 76, 85−86, 90−91, 98, 109−113, 123−125, 127, 128, 133, 137, 141, 161, 163, 166, 172−173, 204−205, 227−228
cautious strategy(ies) 139, 144, 171, 173−174, 178, 215−216, 228, 230
chemical inhibitor 23, 27−28, 33, 36, 53, 75−76, 121, 139, 151−152, 170, 174, 180−182, 186, 192, 206, 216, 225
coleoptile(s) 31
congestion 215
corolla 85, 88, 90
cymbiform seeds (balloons) 89

day neutral plants 64, 136−137, 214

day length(s) 10, 25, 31, 41−43, 45, 50−57, 60−63, 75, 76, 116, 151−152, 191, 198, 204, 210, 212, 213, 223, 225, 226, 229
depression 4−5, 12, 15, 18, 28, 80, 85−87, 89−90, 93−95, 101, 105, 113, 141−143, 147, 149, 176, 179−180, 205, 210, 218, 227
dew 7, 10, 15−17, 85, 92, 118−119, 139, 147, 150, 164, 167, 199, 201, 221, 229
dimorphism 37, 73, 139
dispersal, active 92
− long-distance 26, 86, 144
− mechanism(s) 79, 112−114, 121−122, 140, 142, 171
− passive 82
− seed (see: seed dispersal)
− strategy(ies) 80, 84, 133, 140−141, 143−144, 166, 171−172, 206, 215−216, 228, 230
dispersal unit(s) 15, 29−33, 35−37, 39, 73−74, 76−77, 85, 89−90, 92−94, 105, 138, 145, 147, 158−160, 170, 180−181, 202, 218, 225
− boring 92
− 'creeping' 92
− entangled 90
dormancy (seed) 22, 151, 159, 164, 166, 170, 174, 177, 190, 228−229
drought tolerance 220−221
drying 19, 92, 150
dust-like seeds 81, 84−85, 90, 140
ecogenotype(s) 20, 142, 227
ecotype(s) 14, 20−22, 86, 96, 122, 130, 133, 136−137, 143, 203, 206
embryo(s) 73, 99, 128, 130, 137, 151−154, 158−159, 165, 170, 174, 190, 198, 202−203, 218, 225
endosperm 73, 151, 158−159
environmental effect(s) 10, 36, 41
evaporation 4, 7, 12−13, 175, 182
exozoochory 136, 138, 141

faculative plants (flowering) 214−215
faeces 99−100, 152−153

252 Subject Index

far red light 199−200
female flowers 26, 36−38
fire 182, 199
fruit 25−26, 30−31, 39−42, 45−46, 51,
    53−57, 65−66, 68, 73, 75−77, 85,
    89−90, 100−101, 112−113, 141, 151,
    165, 180, 199, 225−226, 228
generations 25, 36
genoecotype(s) 191
genotypic 23, 25, 74, 76, 145, 150, 170,
    174, 186, 226, 229
germination strategy(ies) 73, 139−140, 144,
    165−166, 171−173, 175, 178, 206,
    215−216, 228, 230
goats 99, 136
grass(es) 14, 84, 92, 177, 197
grazing 96, 136

habitat(s) 5, 12−14, 16, 18, 20−21, 25, 31,
    35−36, 74, 76, 81−82, 95−98, 100−103,
    105, 109, 114−115, 123, 139−140, 142,
    147, 158, 169−171, 174−175, 178, 180,
    183, 189, 194−195, 202, 204, 210, 214,
    216, 217, 221, 226, 230
'hard' seeds (see: seeds 'hard')
harvester ants 95−96, 115
hemicryptophytes 16, 19, 81, 94, 140, 147,
    175, 218
hermaphroditic flowers 26, 36−38
heteroblasty 36, 75
heterocarpy 73, 81, 86, 139
heteromorphism 39, 74
high risk strategy(ies) 171, 178, 215
hilum valve(s) 151−154, 166, 228
hull(s) 33, 36, 180
hydrochastic tissue 118
hydrochory (ombro) 105
hysteranthous geophyte(s) 20

inhibitor(s) (germination) 23, 27−28, 33,
    36, 53, 75−76, 121, 139, 151−152, 170,
    174, 180−182, 186, 192, 206, 216, 225
insect(s) 80, 95, 138

leachate 34, 180
lemma 159
life cycle(s) 12, 18−20, 22, 79, 116, 140,
    164, 169, 173, 175, 198, 210, 212,
    215−216, 223, 225−226, 229
light 22, 25, 27−34, 36−39, 42, 50−52, 55,
    57−66, 68−71, 73, 88, 125, 137−138,
    149−151, 158, 163, 170, 172−173,
    178−180, 182, 184−186, 188−190,
    192−204, 212, 221
light sensitivity 57, 199−200
long day effect 33, 115, 204, 213

low risk strategy(ies) 171, 215

mass germination 23, 76, 105, 138, 140,
    143, 165, 167, 182, 188−189, 199,
    205−206, 221, 226−227, 229
microhabitat(s) 16, 28, 115, 143, 145, 154
mucilage (on seeds) 21, 85, 105, 110, 112,
    128, 130, 135, 147, 221
myxospermy 85, 105, 142, 147−148, 189,
    227

negative photoblasty 202

ombrohydrochory 105
opportunistic strategy(ies) 139−140, 144,
    165, 171, 174, 178, 215−216, 228, 230
Oxygen 130, 137, 164, 190, 203

pappus 26, 39, 73, 86−87, 114, 117,
    119−120, 139, 141, 144
phenotypic effect(s) 25
photoperiod/ism/ic 42, 44, 48, 50, 51,
    54−58, 61, 64, 65, 67, 115, 170, 201, 204,
    206, 212, 229
photothermal effect(s) 41−42
pioneer(s) 37
plant canopy 28, 199
pod dehiscence 110
pollination 142−143, 227
porcupine(s) 4−5, 12, 16, 90, 93−94, 100,
    147, 149, 175, 210, 218−220
porcupine digging 4, 5, 12, 16, 94, 143,
    147, 218
position of capsule 166, 228
 − of flower 25, 36−37
 − of seed 25−26, 29, 35−36, 39−41, 74,
    76−77, 145, 147, 149, 151, 165−166, 170,
    180, 198, 202, 216, 225−226, 228−229
 − on mother plant 25−26, 29, 35−36, 74,
    76, 145, 151, 225

quantitative long day effect 115, 213
 − short day effect 214

rain (rainfall) 1−13, 19−20, 22−23, 26, 28,
    36, 39, 57, 73−74, 76−77, 79−81, 85, 88,
    90−91, 96, 98, 103, 105−106, 109−116,
    118−122, 124−125, 133, 136−143, 145,
    147, 149−150, 152, 154, 158, 164−166,
    169−171, 173−180, 182−183, 185−186,
    188, 190−195, 199, 201, 204−210,
    214−216, 218, 221−223, 225−230
 − distribution 205
 − gauge 28, 174, 176, 180, 182
 − quantities 2

Subject Index

– season 2, 15–16, 18–20, 91, 105, 114, 140, 145, 150, 166, 169, 182–183, 210, 221, 226, 228
red light 33–34, 63, 65–66, 69, 199–201
relative humidity (R.H.) 3, 7, 14, 73, 150, 153–156, 200
rodents 15, 95, 97, 165
roller diaspores 84
rolling plants or inflorescences 90, 97–98, 114
runoff water 4, 93, 95, 105, 120, 141–142, 175–176, 179–180, 218, 227

salinity 4, 7, 21–22, 74, 88, 150, 167, 169, 180, 182–183, 185–186, 206, 225, 229
scarification 43, 46, 53, 152, 166, 228
scarified 21, 125, 128, 153, 154, 155, 170, 183–186
seed(s) (achenes) with pappus 86
– ballist (see: ballist seed)
– bank(s) 15, 18, 26, 29, 36, 47, 51, 60, 76, 79, 81, 100, 103, 111, 115, 125, 142–143, 145, 147, 164, 166, 170, 205–206, 218, 222–223, 226–227, 229
– coat 21, 23, 43, 45–47, 52–53, 62, 75, 91, 130, 141, 148, 151–152, 174, 222, 226
– dehisced by exploding pods 93
– dispersal 12, 23, 76, 79–83, 86–87, 89, 95–98, 102–103, 105–106, 108, 113, 115, 122–123, 125, 133, 137–138, 140, 142–143, 166, 170–171, 188, 203, 205–206 215–216, 218, 222, 226–228, 230
– dormancy (see dormancy, seed)
– eater(s) 14–15, 80, 95, 101, 120, 137, 151, 189, 227
– 'hard' 41, 46, 60, 84, 133, 151–154, 156–158, 166, 170, 218, 228
– maturation 16, 18, 25, 43, 50, 52–54, 57, 61, 63, 66, 70, 74–75, 79, 81, 83, 105, 138, 142–143, 145, 151, 166, 171, 173, 199–200, 205, 212, 214, 223, 225–227, 229
– photoblastic 201–202
– polymorphic 74
– position (see: position of seed)
– predation 79–80, 95, 140, 226
– production/reproduction 80, 84, 95, 143, 165, 169, 173–175, 206, 209, 217, 225
– release by capitula crumbling 110
– size 140
– surface structure 43, 45, 46–48, 95–96, 125, 141
seedling(s) appearance (emergence) 1, 31–33, 74, 95–96, 102, 113, 138, 175, 189, 205, 207–211, 215–216, 221–222, 229

– establishment 7, 28, 138, 166, 170–171, 182, 185, 210, 216, 228
– survival 165, 169, 171, 175, 207, 210, 215–216, 220, 222, 225, 228–229
senescence 41, 47, 57
sheep 136
short day effect(s) 19, 31–33, 47, 116, 152, 213–214
spike(s) 30–32, 36, 89, 90
spikelet(s) 31–34, 36, 89–90, 92, 158–159, 180
spring muturation 60
storage conditions 145, 148, 151, 162, 164, 166, 200, 206, 228
strophiole 151, 157–158
subterranean 'seeds' 15, 28, 41, 60, 143, 152, 157, 216, 221, 227–228
summer maturation 61
summer rain 1, 22–23, 158, 166, 177, 179, 192, 194, 206, 228
synanthus geophytes 19
synaptospermy 85, 90, 105, 142, 148

telechory 142
temperature(s) 1, 3, 7, 10–11, 19–22, 25, 27–28, 30–33, 35, 41, 57, 60–63, 69–71, 74–75, 77, 125, 136–138, 149–152, 157–161, 163–164, 166, 169–170, 173–175, 177–179, 183–186, 188–195, 197–207, 214, 225–226, 228
thermodormancy 158, 190–192
thermoperiodism 195, 206
topochory 142
treasure effect 15, 93, 95
tree(s) 6, 20, 100, 152, 155, 165, 169, 177, 182, 186, 193, 197, 218, 228
tufted seed(s) 86

unopened fruit 85

water stress 37, 42, 71, 73, 136–137, 151, 210, 216, 221, 223, 229
whorls 36, 116, 118, 121
wind 26–27, 29, 73, 80–82, 84–96, 102, 104, 110–115, 120–122, 138–141, 144, 147, 149, 152, 203, 226
wind trap(s) 93
winged diaspore(s) 85, 90
winter annual(s) 1, 10, 18–19, 22, 39, 90, 141, 143, 158, 163, 177, 179, 190–191, 204, 206–207, 210, 214, 218, 226
winter maturation 10, 19–20, 25, 39, 51, 60–61, 64–67, 71, 81, 105, 110, 121, 138, 141, 143, 165, 204, 214, 226
winter rain 1, 22, 192, 194, 204, 206